essentials

Chirine Etezadzadeh

Smart City – Future City?

Smart City 2.0 as a Livable City and Future Market

Dr. Chirine Etezadzadeh
Ludwigsburg
Germany

ISSN 2197-6708 ISSN 2197-6716 (electronic)
essentials
ISBN 978-3-658-11016-1 ISBN 978-3-658-11017-8 (eBook)
DOI 10.1007/978-3-658-11017-8

Library of Congress Control Number: 2015949088

Springer Vieweg

Printed on acid-free paper

Springer Fachmedien Wiesbaden is part of Springer Science+Business Media
(www.springer.com)

What you Can Find in this *Essential*

- a holistic view of cities as urban systems representing a complex organism
- a description of social developments in cities based on global mega trends
- a discussion on the topic of digitalization including a future outlook
- a deeper understanding of the concept of a smart city
- a broader definition of sustainability and an urban design approach including ideas and implications for product development

Definition of "City"

"A city as opposed to the country or rural areas is a larger, densely populated settlement that has specific functions in terms of the geographical division of labor and political power, depending on the social organization and types of production. In the Federal Republic of Germany, for instance, an urban settlement is defined as a chartered municipality with a population of 2000 or more according to official statistics (a rural town has a population of 2000–5000, a provincial town a population of 5000–20,000, a medium-sized town a population of 20,000–100,000, and a large town/city a population of more than 100,000)."[Author's transl.]

Source: Gabler Wirtschaftslexikon (n.d.). keyword: city. http://wirtschaftslexikon.gabler.de/Archiv/9180/stadt-v9.html. Retrieved on 01/01/2015.

Preface

Whenever the topic of smart cities arises, the discussion quickly turns to images of otherworldly engineering ingenuity or visions of IT-based omnipotence. This *essential* aims to provide a context for those future scenarios. It deals with cities as places of action not only as relevant markets of the future, but also as living spaces. The analysis presented in this publication thus provides insights into a range of topics regarding cities of the future, on the one hand, but can also serve as a basis for market-oriented and customer-centered urban product development, on the other.

Although it is compact, this *essential* provides a comprehensive overview of the very dynamic global urban situation. Its purpose is to spark interest and point out that we all need to focus our attention on cities and current social trends. This publication is based on Dr. Chirine Etezadzadeh's consulting and research work and on her series of lectures on "product development for smart cities." More information is available at **www.SmartCity.institute**.

Contents

About the author

Dr. Chirine Etezadzadeh (economist) is a strategy consultant who deals with the future on a daily basis. Her work and research focus on business development and customer-centric product development. As a specialist for the automotive industry and the energy sector, she worked in the fields of future and electro mobility. This led her to explore the topic of smart cities, which has become one of her major areas of expertise. Today, Dr. Etezadzadeh holds university lectures on product development for smart cities to help pass knowledge on to the next generation.

For several years, Dr. Etezadzadeh worked for a German premium car manufacturer, a leading U.S. automotive supplier, and as a business consultant for the energy industry, besides regularly engaging in interdisciplinary research activities. In 2009, she founded the consulting firm THINK and GROW consult. Since that time, she has been providing consulting services to industrial corporations and cities as well as to small and medium-sized companies regarding strategic matters. In the summer of 2014, she launched the SmartCity.institute as a platform for research on future cities.

Introduction 1

Cities are important. They are increasing in number and expanding; they account for a major share of global economic output, are gradually gaining more political influence, and can leverage action to combat climate change and protect the environment. Cities are markets whose overall needs should be identified and incorporated in solutions. Cities are above all living spaces—for a steadily growing number of people. That is why they should be designed to provide a sustainable and livable environment for all of their residents.

A city is comparable to a human organism in which many substances, agents, and processes interact to keep the body functional. A city needs to be nourished, cleaned, cultivated, and nurtured in order to thrive and grow. The care and attention it receives will give it self-confidence and protect it from different types of threats. As it matures, a city becomes capable of identifying its needs and finding ways to satisfy them, of being creative and generating new life. In the following, we will develop a deeper understanding of what characterizes a healthy city of the future.

C. Etezadzadeh, *Smart City – Future City?*, essentials,
DOI 10.1007/978-3-658-11017-8_1

The City as a Place of Opportunity 2

2.1 Challenges Facing Cities

What do people expect when they move to a city? As diverse as their motives may be, they always have one element in common: *access*. Access to jobs, a livelihood, and perhaps even affluence,—market access; access to the necessities of life, such as water, food, housing, and health care services; access to infrastructures: to electricity, heating, sanitary facilities, waste disposal systems, etc.; access to information, knowledge, technological advances, and—with a lot of luck—education; access to other people, to a social, cultural, or religious life, to special groups or like-minded people and communities, or to anonymity; access to a place where people have rights and obligations conftrolled by institutions and defended by courts, if necessary; a place that offers a certain degree of security, stability, and predictability, including protection against threats such as natural or man-made disasters. Cities provide access to opportunities, to physical and potential social mobility, and in varying degrees to the big, wide world (e.g., via products, a movie theater, customer/supplier relations, tourists, or through a train station or an airport). The desire to seek access is apparently one of the reasons why urban populations are rapidly growing worldwide. Today, over 50 % of the world's population lives in cities, a proportion that the United Nations (U.N.) expects to increase to 66 % in 2050.[1]

Even though the idea of "hoping to gain access" is often due to a lack of prospects for the future or, at the other end of the scale, due to superfluity, boredom, or a concentration of power, it naturally also implies the search. To be more specific, the search for access that could become attainable, which becomes the goal and needs to be earned and maintained at best. The search thus includes

[1] cf. UN/DESA (2014, p. 2).

C. Etezadzadeh, *Smart City – Future City?*, essentials,
DOI 10.1007/978-3-658-11017-8_2

the endeavor. And that endeavor represents the most fundamental form of urban dynamics—namely, that of the city dwellers. They seek self-preservation and self-fulfillment, or perhaps they seek to secure a livelihood for their families or wish to develop and improve their own living conditions or those of others. Yet they are not alone in pursuing those goals. Due to the population density in cities, urbanites face increased competitive pressure in all their endeavors. This aspect drives urbanites to pursue their endeavors more intensely, thus considerably accelerating urban dynamics. That dynamism gives rise to innovation that, depending on the overall situation, can have positive results, such as enhanced services and benefits, or negative results, such as more unscrupulous types of crime.

Cities are constantly changing. They are centers of learning and development. Thus, the first industrial revolution—the introduction of mechanical production processes, the use of steam engines, and the use of coal as an energy source—took place in cities located in countries that later became industrial nations. The second industrial revolution—automated and mass production in conjunction with the introduction of a centralized power supply and electrification[2]—also originated in cities; over time it has shifted, together with its inherent problems (environmental and social aspects), from the industrial nations to cities in less developed regions of the world. Advances in communication media ranging from telegraphy to telephones, radios, television, and the Internet also initially occurred in cities. And the "third industrial revolution," the digital revolution, which is already under way, likewise began in cities: this includes (a) digitalization, (b) the virtual decentralization of production processes and our lifeworld, as well as (c) key stimuli for an energy transition (although in Germany the energy transition actually did have its origins in rural areas). It is not yet possible to determine which cities and regions will be the leaders in the third industrial revolution. However, the leadership roles will manifest themselves within the next few years.

Not only the city inhabitants and their government form the basis for developments of this kind, but also initially the natural environment in which a city is embedded. Whereas processes in nature are inherently a type of cyclical economy based on "life—death/decay—recycling/return to life," this does not hold true for human production and consumption. This becomes vividly apparent when a city is viewed as an urban system. By overusing the available resources, urbanites exploit their natural environment, thereby destroying it and endangering their own basis of existence. As major polluters, cities foster climate change, the consequences of which they are particularly exposed to.[3] At the same time, however, due to their

[2] cf. Schott, D. (2006, p. 255).

[3] cf. Revi, A., Satterthwaite, D. E. (2014).

density and structure cities have the potential to manage their economic activities in an environmentally friendly and resource-efficient manner and to take suitable action to promote the protection of their natural environment. We need to quickly tap this potential.

Progressive urbanization, the third industrial revolution (including the obligatory energy transition), and above all the preservation and cultivation of their natural environment and thus their basis of existence are three major challenges that cities face. While cities in many parts of the world are growing, mature cities in aging and already urbanized industrial nations are stagnating or even shrinking in size. In Germany, for instance, especially cities in rural or economically underdeveloped and structurally weak regions have to contend with shrinking populations, or with the combination of growth and shrinkage. Urban development in industrial nations is essentially characterized by demographic trends in terms of a shrinking and aging population, the latter being a global phenomenon.[4] In a broader sense, demographic change also includes aspects such as rising numbers of immigrants (heterogeneity) and a diversification of lifestyles (e.g., singularization/single-person households), two trends that likewise distinguish urban life worldwide.

2.2 The Relevance of Cities

What are the reasons for the progressive *urbanization* in terms of "the expansion and intensification of urban lifestyles, economic activity, and culture" [author's transl.][5] that is taking place? First of all, we need to note that the world population is rapidly increasing. In July 2013 the world population was 7.2 billion. Between 2005 and 2013 it rose annually by a figure approximately equal to the population of Germany. Taking into account the assumption that fertility rates will continue to drop, the U.N. nevertheless expects the world's population to increase to 9.5 billion by 2050 and to 10.9 billion by 2100.[6] Africa is projected to have the highest growth rates between 2011 and 2030. During that same period, Asia is expected to have

[4] cf. UN/DESA (2013b, p. xviii ff.).

[5] Bähr, J. (2011a, n.p.).

[6] cf. UN/DESA (2013b, p. xviii).

the highest population growth by far in absolute numbers. The U.N. reports that the world population is growing especially in the developing countries.[7,8]

An increasing proportion of this growing population lives in cities. Whereas the urbanization rate in the industrial nations (developed countries) is already almost 80 % today and is expected to rise to 85.9 % by 2050, it is projected to increase from 46.5 % in 2011 to 64 % in 2050 in the less developed (high growth rate) regions.[9]

The following examples serve to help illustrate these numbers. Let us take a look at India. According to U.N. estimates, an average of 21 people living in rural areas in India will move to metropolitan areas every minute within the next 20 years. In order to accommodate that massive inflow of people to urban areas India will need about 500 new cities during that same period. Between 2014 and 2050, India's urban population will increase by 404 million people and China's by 202 million (according to U.N. estimates).[10] In Bangladesh the population of Dhaka grew by 3259 % between 1955 and 2015 to 17.6 million.[11] These numbers defy the imagination of Europeans and present local administrative bodies with tasks that are practically impossible to accomplish.

Apart from (a) natural population growth, the reasons for progressive *urbanization* in terms of "the proliferation, spread, or growth of cities by number, surface area, or population both in absolute numbers and in relation to the rural population or nonurban settlements" [author's transl.][12] are (b) migration from rural areas, and (c) the urbanization of rural areas, the latter—creation and reclassification of cities—tending to be a less significant factor.[13]

As a result of these trends, cities such as Tokyo with a population of 38 million and Delhi with a population of 25 million as well as sprawling urban zones with even larger populations already exist today. The number of megacities (cities with a population of more than 10 million) has increased considerably. In 1970 only two

[7] "More developed regions comprise all regions of Europe plus Northern America, Australia/New Zealand and Japan. Less developed regions comprise all regions of Africa, Asia (excluding Japan), and Latin America and the Caribbean as well as Melanesia, Micronesia and Polynesia. Countries or areas in the more developed regions are designated as 'developed countries'. Countries or areas in the less developed regions are designated as 'developing countries'." Source: UN/DESA (2013b, p. vii).

[8] cf. UN/DESA (2013b, p. xix).

[9] cf. UN/DESA (2012, p. 4).

[10] cf. UN/DESA (2014, p. 1).

[11] cf. UN/DESA (2002 and 2014).

[12] Bähr, J. (2011a, n.p.).

[13] cf. Bähr, J. (2011b, n.p.).

megacities existed, whereas today there are already 28, a number that is expected to rise to 41 by 2030. Today, 12 % of the urban population lives in such metropolitan agglomerations.[14] Despite the steep increase in the number of megacities, they are not an urbanite's first choice. The majority (almost 50 %) of the world's urban population lives in cities with a population of less than 0.5 million.[15]

The ongoing urbanization process naturally has a dramatic impact on the environment. To give a few examples, we could mention aspects such as the use of land (e.g., for transportation or housing), the use of natural resources (e.g., for water and energy supply systems), the contamination of natural resources (e.g., soil and water), air pollution (due to manufacturing, heating of buildings, traffic, etc.), and noise emissions. Urbanization often leads to an irreversible depletion of our natural resources, to the destruction of natural environments, and to a loss of biodiversity. Urban planning therefore needs to be aimed at creating compact cities whose economies use resources efficiently in order to ensure the viability of the city and its environment. This objective appears utopian in view of the many metropolises that routinely grow uncontrollably at random, yet it is crucial.

In brief it can be said that, irrespective of where in the world they are located and regardless of their size and underlying circumstances, what all cities have in common is the obligation to identify the needs of their inhabitants, adapt to the increasing demands placed on them, and transform themselves accordingly. The situation described above suffices to demonstrate the relevance of cities. Cities will shape the future and are markets of the future.

2.3 Goals for Cities

Once it has been acknowledged that a healthy environment, people's quality of life, and economic viability are inextricably interlinked and interdependent, urban developers generally try to incorporate the *principles of sustainability* in their urban planning projects. "Sustainable development means that environmental aspects and social and economic aspects receive equal consideration[16] and that the needs of today's generation are met without jeopardizing the ability of future generations to satisfy their own needs and select their own lifestyle."[author's transl.][17]

[14] cf. UN/DESA (2014, p. 1).

[15] cf. UN/DESA (2014, p. 1).

[16] cf. Rat für Nachhaltige Entwicklung (n.d.), n.p.

[17] Lexikon der Nachhaltigkeit (n.d.), n.p.

Based on those principles, common goals emerge that many different types of cities focus on to a greater or lesser extent, depending on the underlying circumstances. These goals include:

1. protecting (in the interest of preserving) the natural environment, the climate, and resources, i.e., the urban living conditions,
2. maintaining the urbanites' quality of life, or promoting the social development of the city, and
3. maintaining the city's competitiveness, or promoting its economic development,
4. … for current and future generations.

None of these goals can be kept alive without taking the others into consideration. Consequently, in a healthy city none of these goals may be neglected. In the following, we will refer to sustainability requirements as urban *"meta-goals."*

But what is it like in cities in reality? The initial years of the new millennium were generally characterized by economic stagnation and recession primarily due to the prolonged financial and economic crisis. Even emerging markets are currently no longer growing at the rates they achieved in the past. It can furthermore be noted that economic growth no longer necessarily entails job creation and social progress.[18] The increasing decoupling of capital growth from productivity has resulted in a more rapid accumulation of wealth and greater poverty. Due to structural changes, a large portion of the population has been driven out of the labor market or forced to accept low-skilled and low-paying jobs and/or precarious employment.[19] This part of the workforce has lost its middle-class status (in industrial nations) or has started to work in the informal sector, which accounts for a significant share of the economic output in less developed countries. Apart from all the anguish associated with the informal economy, it does in part enable urban life in places where state or municipal support is unavailable.

When people lose their means of existence (e.g., in the agricultural or industrial sectors, or because of climate change, terrorism, or war), fear, poverty, a lack of prospects for the future, and unemployment lead to international migration or to internal migration from rural to urban areas. In cities, random, uncontrollable urban growth results in inner-city segregation, i.e., in social and geographical division. This trend adversely impacts a city's "manageability," lowers the quality of life of the city's residents, encourages environmentally detrimental behavior, and jeopardizes the city's economic success because of the enormous subsequent costs.

[18] cf. COM GD REGIO (2011, p. VI).

[19] cf. COM GD REGIO (2011, p. VI).

This is a self-accelerating, systemic process that has the potential to cause cities to deteriorate and turn into uninhabitable behemoths or to give rise to no-go areas where people do not feel safe. In this context, the question arises as to what happens to people who—in hopes of gaining access—leave their homes to move to a city to find work, but end up without any prospects for the future or are mercilessly exploited.

In order to call attention to these kinds of interdependencies that enable inhuman and environmentally destructive production processes on which modern consumption is based and to point out that due to globalization "future consumers" in developing countries will also want everything considered to be standard and a status symbol in Western industrial nations, we have added another dimension to the sustainability requirements. This dimension defines a requirement that can hardly be met but, just like the other three dimensions, could set a trend and serve to raise awareness regarding human consumption patterns. The author is of the opinion that the sustainability requirements are no longer adequate in view of globalization, an increasing focus on economic aspects, and the progressive destruction of the environment. Therefore, the requirement to be *generalizable* has been added to the municipal meta-goal system as shown in Fig. 2.1. This means that approaches, decisions, or actions and their consequences should be assessed as to whether they are still sustainable and tenable if they are repeated in different contexts by different and/or numerous stakeholders.

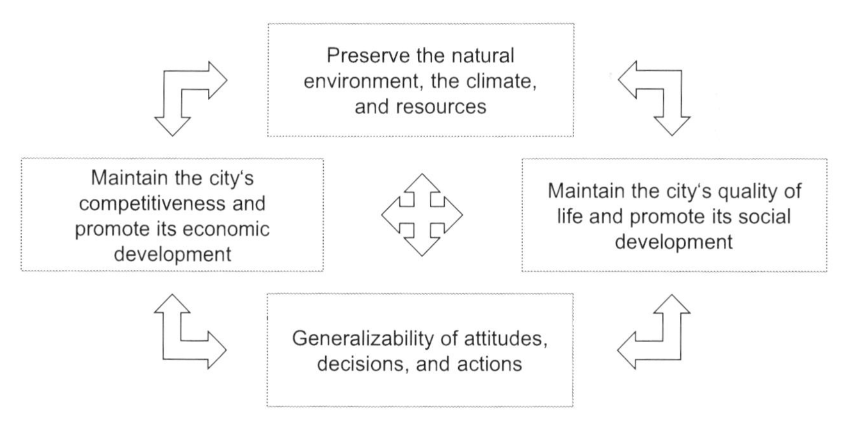

Fig. 2.1 The municipal meta-goal system, including the requirement for generalizability. (Source: author's own graphics)

Taking the above into account, a city should pursue a clear, (if possible) participatory written, individual *vision* to which all urbanites and stakeholders feel committed. It should include all residents and their basic necessities as well as the preservation and cultivation of the living and natural urban environment and should give consideration to both global sustainability goals and the city's specific underlying circumstances.—However, the top municipal goal uniting all cities has not yet been addressed.

3 Functionality

3.1 Infrastructures

The overriding municipal goal is a city's *functionality*. Functionality is largely determined by infrastructures. They enable urban coexistence and the pursuit of meta-goals. Infrastructures dictate the living and working conditions of the urbanites.

In the following, we will discuss mainly the technical infrastructures of cities. Certain social infrastructures are an exception: the food supply, health care, education, and services. Technical infrastructures (referred to below as *infrastructures* for purposes of simplification) are significant factors in tackling the most basic urban challenges, important fields of research, and relevant markets of the future for the industrial sector. Figure 3.1 presents an overview of the sectors we have taken into consideration together with the municipal meta-goals.

Infrastructures should be regularly and systematically overhauled or modernized in accordance with relevant urban needs in order to

a. ensure that residents have continuous and unhampered access to utility services,
b. be able to provide the mostly public goods at locally reasonable prices or free of charge,
c. keep up with the technical advances that make economic, social, and environmentally responsible life that meets modern standards possible.

Depending on their condition, the facilities need to be adapted to meet capacity requirements (investment in expansion, dismantling, and removal), repaired (investment in replacement/modernization), and/or renovated or "smartly" retrofitted so that they meet contemporary standards and needs (investment in adaptation). Projects of this type are very challenging because of the nature of the investments (extended planning periods), the long service life (capital commitment), and the

C. Etezadzadeh, *Smart City – Future City?*, essentials,
DOI 10.1007/978-3-658-11017-8_3

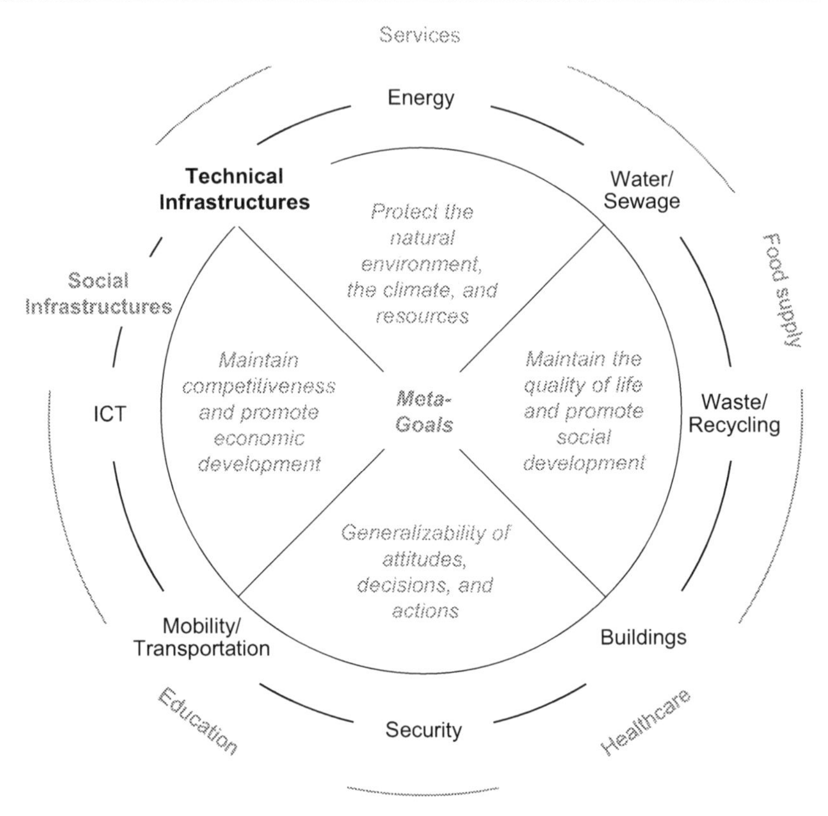

Fig. 3.1 Overview of the technical and social infrastructure sectors discussed in this chapter and municipal meta-goals. (Source: author's own graphics)

large amount of capital required. Moreover, infrastructures cannot be divided into smaller projects (minimum size/performance requirements that do not exactly match demand).[1] This situation is due in particular to current financial and economic difficulties and the persistent demand made by urbanites to be allowed to actively participate in urban planning processes.

A look at the status of the infrastructure sector vis-à-vis the actual needs is disconcerting.[2] Emerging cities with high growth rates like Lagos, Karachi, or Mumbai often lack even the most basic infrastructures. Growing, transitional cities that continue to attract newcomers, such as Beijing, Seoul, or Istanbul, frequently lack the required capacity in the infrastructure sector, and some of their facilities

[1] cf. Gabler Wirtschaftslexikon (n.d.), n.p.

[2] cf. Siemens (2006).

are in part outdated. Mature cities whose populations tend to be stagnant, such as Tokyo, New York City, or London, usually have infrastructures that have existed since the beginning of the last century or even longer. Many of the utility installations in these cities need to be repaired and modernized, and some of them are oversized and inefficient.[3] Industrial corporations and consulting firms see a great need for infrastructure modifications and regard this sector as a future market to be developed.[4]

The Consequences of a Lack of Infrastructures

Delaying measures to regularly build or improve infrastructures entails high subsequent costs. Utility installations that become partially or even wholly inaccessible have social consequences resulting in costs that are difficult to quantify. The author would like to present three examples below[5]:

Informal settlements without any or with insufficient access to infrastructures encourage the manipulation of utility networks, illegal dumping, the burning of toxic materials and fuel, contamination of the living environment, the lack of hygienic standards, and the consumption of contaminated water. Diseases due to contaminated water or an unhygienic environment lead to high rates of mortality and entail the risk of epidemics in a society that lacks health care and is often malnourished.

The lack of a public mobility infrastructure can lead to the emergence of massive, partially uncontrolled private motorized transport, resulting in traffic jams (i.e., social costs/economic losses), traffic accidents, air pollution, noise, etc. A lack of access to public transportation in urban areas (e.g., informal settlements) leads to social exclusion and to the escalation of precarious developments.

Lack of access to mobility infrastructures and a lack of access to jobs or employment providing a livelihood foster the emergence of an informal sector, precarious conditions, and potentially criminal activity, terror, and organized crime. The lack of educational and care facilities exacerbates these developments. Urban socio-economic disparities arise. Inequality and a high crime rate instill fear. Social segregation ensues:"gated communities" emerge that reinforce the process of exclusion and prevent access. Housing becomes more expensive. Affordable housing is scarce. This leads to suburbanization and the formation of new informal settlements.

The consequences of a lack of infrastructures are complex, extremely interdependent, and affect all municipal meta-goals. They show that linear thinking is not adequate for urban development. The intricacies require integrative, systemic

[3] cf. Siemens (2006).

[4] cf. ZVEI (2010).

[5] cf. COM DG REGIO (2011).

thinking and action. That is why a wide range of professional and methodological expertise is necessary in the field of infrastructures to allow intersectoral and interdisciplinary thinking and action.

3.2 Critical Infrastructures

Some infrastructure sectors are especially relevant with respect to the functionality of cities. Referred to as *critical infrastructures*, these are "organizations and facilities that are important for the national (or urban [the author]) community, the loss or impairment of which would cause prolonged supply shortages or a considerable disruption of public safety, or have other dramatic effects." [author's transl.][6]

Infrastructures display significant intersectoral interdependencies that extend far beyond purely technical infrastructures. Particularly in the area of critical infrastructures, disruptions can therefore have domino and cascade effects. They have the potential to bring parts of society to a standstill, can cause great economic damage on top of the damage directly incurred by the people concerned, and can lead to a loss of confidence in the political leadership of a society.[7] As an example, we would like to describe an accident that occurred in Baltimore as follows:

On July 19, 2001, a freight train transporting chemicals derailed in an urban tunnel. As was to be expected, this caused damage to rail and road traffic and required the help of the appropriate emergency services. In addition, the incident also had unexpected consequences. For instance, the fire in the tunnel caused a water pipe to burst, which led to flooding up to a level of 90 cm in some areas besides producing 6-m high geysers. Twelve hundred city residents experienced a power outage because of the accident. Damage to fiberglass cables caused widespread disruptions in landline and cell phone communication and Internet services. Data service operations at several corporations failed in the area of critical infrastructures, among others. Significant disruptions in rail traffic affected neighboring states. The effects included disruptions in the supply of coal and raw materials for steel production, for example.[8]

Cities and societies are vulnerable. Terrorist attacks and accidents due to technical failure or human error remind us of this, as do natural disasters that are occur-

---

[6] BMI (2009), p. 4.

[7] cf. BMI (2009), p. 9.

[8] cf. Pederson et al. (2006), p. 4.

ring more frequently nowadays, which is believed to be associated with climate change.[9] This gives rise to the phenomenon described below:

> As a country's utility services become less prone to failure, the greater the effects of any disruption will be (…). In the course of their technological development, societies begin to react much more sensitively to disruptions, especially when they involve infrastructures based on advanced technologies, since the public is used to having very high safety standards and a high degree of security of supply. This situation in which people have developed a rather deceptive feeling of security as services become increasing reliable and less prone to failure, although the consequences of any 'incident that happens to occur anyway' are disproportionately greater than before, is referred to as the *vulnerability paradox*. [author's transl.][10]

In other words, the higher our standards are, the more vulnerable we become because of our expectations. We need to consider this aspect when designing a smart city.

3.3 Resilience

In view of the above, the German Federal Ministry of the Interior has called for a new risk culture as society faces growing vulnerabilities. The proposed risk culture would include more open communication concerning risks, better cooperation between the stakeholders, more voluntary commitment by the operators of critical systems (often private sector companies), as well as greater self-reliance and increased self-help efforts on the part of the people and facilities affected by incidents.[11] Or to put it a different way and with reference to our field of research, what that means is that cities need to become more resilient. "*Resilience* is the ability to prevent actual or potentially adverse events from occurring, to prepare oneself for them, take them into account, cope with them, recover from them, and adapt to them more and more successfully." [author's transl.][12] Natural disasters, technical failures or human errors, terror or war—which is easy to conduct nowadays by manipulating critical infrastructures (cyber-attacks)—and simply the wear and tear on facilities all pose a threat.

[9] cf. BMI (2009), p. 9.

[10] BMI (2009), p. 11 ff.

[11] cf. BMI (2009), p. 11 ff.

[12] Acatech (n.d.), n.p.

Resilience is a key aspect of sustainability because its purpose is to preserve the functionality of a city, i.e., to enable urban life on a "sustainable" basis and help achieve the city's meta-goals. Resilience requires the creation of robust, fault-tolerant, reliable infrastructures[13] and sophisticated emergency plans that take the increasing interdependencies of infrastructures and potential cascade effects into account. It furthermore requires cities to be self-sufficient to a certain degree, which entails urban production and a local supply of food. Ultimately, it requires urbanites to assume a new attitude. They should be flexible and adaptive in accepting change and incorporate resilience in their planning processes and actions as a matter of course.

Cities need to raise the awareness of their stakeholders and residents for those interdependencies and encourage them to adapt their behavior as necessary. As opposed to countries, the majority of cities encompass a manageable area where people's attitudes can be encouraged and developed. This can be achieved by helping urbanites to understand the extent to which each one of them can benefit from pursuing their common goals. Aspects such as sustainability, an appreciation for preservation, cleanliness, responsibility, community spirit, a willingness to help, solidarity, civil courage, etc. that boost resilience could be incorporated in the urban attitude towards life. In addition, urbanites could be encouraged to assume more responsibility by institutionalizing this role and offering incentives to a greater extent than is currently the case. Providing for and helping one another in default/emergency situations should be as natural in a city as the carnival in Rio, earthquake drills in California, or cleanliness in Singapore, though the latter is mostly imposed by the government. A coffee-to-go society, however, needs 24/7 services. It will have trouble coping with emergencies.

Cultural change is a project requiring comprehensive efforts to get everyone onboard. Yet it is necessary and appears to be feasible in an urban context. Along these lines, the city of Cologne's motto, "love your town," which its residents have embraced, could become the global slogan for an urban resilience campaign.

[13] cf. Acatech (n.d.)

4 Urban Product Requirements

4.1 Urban Consumption and Ideal Products

Cities whose inhabitants pursue sustainable visions and goals, cities that overhaul their infrastructures in order to remain functional and want to become resilient by initiating a cultural change will also place new demands on their products. Products characterize urban life. This is where they are developed, produced, sold, used, consumed, disposed of, and recycled. However, limited resources, the need for entrepreneurship and jobs, a lack of space, piles of garbage, water, air, and soil pollution due to production, transport, and consumption, and noise emissions are at least currently still a part of urban reality. What implications does this have for the development and design of products for cities?

Let us first define the meaning of the term *product* as it is used in the following. In marketing, a product is anything that can be sold. We would like to narrow the definition of product slightly, not to take the edge off the above statement, but rather to better illustrate the subject matter to be discussed. For the sake of simplicity, let us take a look at material goods—consumer goods and capital goods.

In a sustainable city, it makes sense to use products that do not have any negative effects on the environment, climate, or resources, and are beneficial for the quality of life and social progress, while keeping the economy going at the same time; products that form the basis of business models that safeguard jobs and are tailored to meet the specific needs of a city. That would give homo sapiens urbanus a relatively limited number of consumption options. At any rate, the majority of products do not conform to these ideas.

In light of the above, a product process that focuses on the expanded notion of sustainability described above and on resilience is a targeted approach and expedient businesswise as well, considering that social trends towards more sustainable

C. Etezadzadeh, *Smart City – Future City?*, essentials,
DOI 10.1007/978-3-658-11017-8_4

consumption are emerging. This approach, which we will call an *"urban design approach,"* is geared to eco-design and cradle-to-cradle design, giving consideration to and forming the entire product cycle in accordance with the requirements for sustainability and resilience. The product cycle ranges from the product development process and the sourcing of raw materials, the product creation process, all transport routes, product marketing, and the product's useful life, all the way to the disposal and recycling of the product and its constituent materials (please refer to Fig. 4.1). The purpose of this design approach is to create products that are incorporated in materials cycles oriented towards natural, waste-free processes.

The urban product process begins with a product development process during which all stages of the product cycle are thoughtfully designed in such a manner as to prevent any negative external effects. Thus, product development is based on the use of "clean" (mining and production) and fair-trade raw materials derived from renewable resources. With respect to animal products, this means largely avoiding the use of animals, species-appropriate animal husbandry, and more animal-friend-

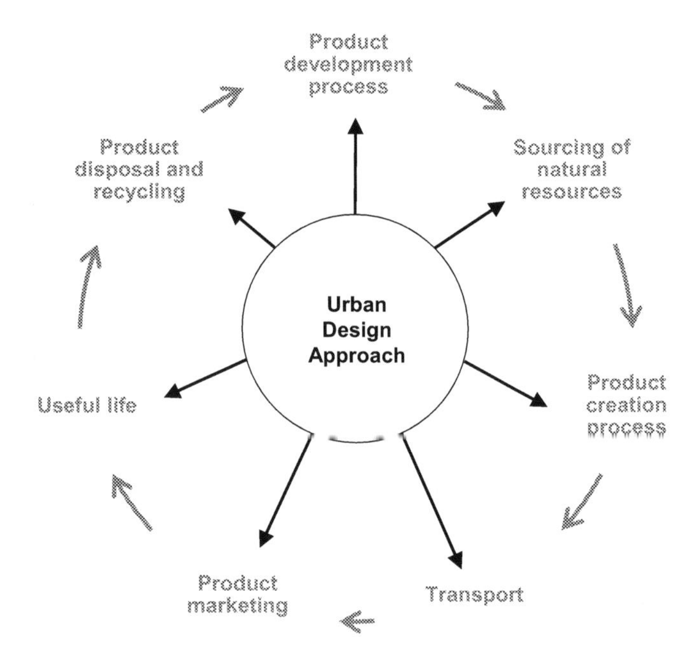

Fig. 4.1 The product cycle of the urban design approach. (Source: author's own graphics)

ly slaughtering than is currently the practice in factory farming. The entire power supply is produced from renewable energy sources.

The product creation process is designed to be sustainable, taking into account the requirement to minimize transport distances (e.g., through urban production instead of global supply chains) that applies to each product stage. Collaboration with upstream and downstream businesses and staff is characterized by stability, fairness, and working in partnership.

Marketing aimed at avoiding waste and preserving value may perhaps be based on new business models (e.g., leasing, sharing, contracting, or on-demand services, etc.), since manufacturers think in terms of new value creation units (e.g., mobility vs. automobiles) and service structures (e.g., as energy service providers).

The useful life of the products is designed for durability and resilience. This not only means that a product satisfies customer needs, but also that the product itself displays attributes of a long service life.

Plans should also be in place for the final stage of the product life cycle. The ability to take a product apart and to recycle and reuse its materials is something that needs to be planned step-by-step and well in advance (e.g., by anticipating future recycling processes) in order to consistently eliminate standard "recycling processes" like those in Agbogbloshie.[1]

The product process outlined above is defined during the product development stage, as previously mentioned. During that stage, the details of every stage of the cycle are laid down (except with respect to marketing). That is why the product development stage is the key element in the urban design process.

4.2 Urban Product Development

Let us consider product development from a slightly abstract perspective. During the development stage, the *whole product (Produktgesamtheit)*, which consists of three *core attributes (Wesensmerkmale)* of products, basically needs to be designed end-to-end and tailored to meet customer requirements.[2] Accordingly, a product's (a) instrumentality, (b) materiality, and (c) semiocity (symbolic character) as well as the resultant *objective character (Gegenständlichkeit)* must be thoughtfully developed and designed during the entire product process. This will allow the creation of products that are inherently compelling and successful (refer to Fig. 4.2).

[1] cf. WDR (2012). English video: cf. SBS Dateline (2011).

[2] A detailed description of this product theory can be found in Etezadzadeh (2008).

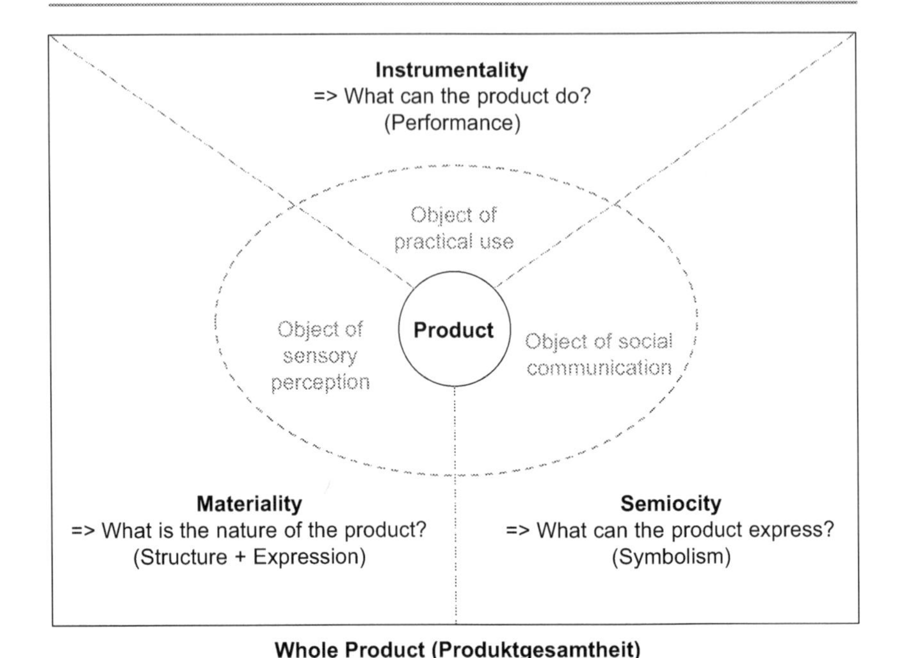

Fig. 4.2 The three core attributes of products and their objective character. (Source: Based on Etezadzadeh (2008), p. 39)

Instrumentality stands for what a product can do and addresses its performance. Materiality denotes the nature of a product and addresses its structure and expression. And semiocity (symbolic character) refers to what a product can express and stands for its symbolism.[3]

The urban product process as envisioned above places additional demands on the design of these essential features. Consequently, urban products must be designed to be sustainable and resilient at the very least. The products must be appealing and durable in order to be sustainable during the product cycle. Products of this kind justify product-related investments, the commitment of human resources and capital, the use of natural resources and materials, etc., and aspects such as path dependencies that may be associated with the production process, the use of technology, or societal decisions in favor of products.

As regards resilience, it is necessary to create products that have minimum downtimes, "continue to function or can automatically switch to a safe state even

[3] cf. Etezadzadeh (2008), p. 39.

in the event of disruptions,"[author's transl.][4] and are safeguarded against urban risks (including espionage, cyber-attacks, and terror). Research institutions therefore call for "resilience by design," i.e., "the integration of system resilience during the initial stages of the design process, making resilience a fundamental prerequisite for any technological or societal security solution."[author's transl.][5]

What implications do the requirements for sustainability and resilience have for the design of the three product dimensions?

A product as an object serving a practical purpose (instrumentality) must primarily offer benefits and, as described above, satisfy a persistent customer need over the long term.[6] In order to extend a product's service life and its period of usability it should be maintainable, easy to service, reparable, durable, adaptable, innovation-friendly, capable of being updated, interoperable, etc., and meet contemporary requirements (e.g., for use in sharing systems)—insofar as it can have these features. In view of the social developments described in Sect.. 2.1, it should furthermore be barrier-free, easy to handle, integrative, and ease burdens.

From the manufacturers' perspective, products for urban areas should additionally have a scalable and modular design and their total cost of ownership (TCO) should be as low as possible. On the one hand, this would allow a basic version of the products to function and be marketed in various contexts. On the other, it would be possible to offer a wide variety of options since the products could be adapted to meet different needs, existing dimensions, technological standards, changing circumstances, and available municipal budgets without becoming too complex. Moreover, this would serve to bolster the products' resilience and sustainability.

It becomes clear that the *materiality* of a product is closely intertwined with its *instrumentality*. It enables the product's long-term use and turns the product into an object of sensual perception. In combination with the materiality, the attributes mentioned above that help make products durable are designed to permanently improve both a product's longevity and its appeal and to make the product enjoyable to use (experience aspect). The material is the direct point of contact with the user and thus forms the basis for a lasting product relationship. Maximum benefit can be derived through long-term use of the product and thoughtful recycling of the reusable materials by choosing a material that is appropriate for the product and designing the product with due regard to material and recycling aspects. The selection of materials to be used should not include nonrenewable resources, toxic materials, or materials that have adverse health effects. With respect to sustain-

[4] Acatech (n.d.), n.p.

[5] Acatech (n.d.), n.p.

[6] This generally holds true for capital goods, but not necessarily for consumer goods.

ability and resilience, the goal is to achieve a combination of durability and a high level of appeal.

The *symbolic character* of products is likewise closely intertwined with the other two attributes. It appears to be somewhat less significant for capital goods than for consumer goods, but is still of utmost relevance. For instance, the most modern technical facilities whose symbolic character is based on technological visions of the 1960s would not be very promising from a business perspective.

As regards the symbolic character of a product, the question arises as to what needs to be done to make a product's symbolism sustainable and resilient and how to communicate these two aspects. First of all, this can be more readily achieved by using a design that is not too trendy (sustainable) or too context-related, i.e., the product's symbolism should be easy to understand (resilient).

In order to communicate the fact that a product meets sustainability requirements its symbolic character could, for instance, signal its intrinsic value and durability or certain sustainable features, thereby serving to enhance the prevailing status symbolism. To give one example of this: In the automotive industry, sustainability can be associated with lightweight design, which is reflected in automobiles featuring lightweight yet high-strength structures, for instance. Those kinds of features can be stage-managed or metaphorized by using appropriate designs or suitable symbolism. Accordingly, a car's interior could be designed to appear more delicate, or "lighter," while remaining sturdy and safe. In the history of the automobile, there are numerous examples of a symbolic character sometimes used rather boldly to signify technological sophistication or specific performance features.

And how do you communicate resilience? By stage-managing mechanics or by using a sturdy design or heavy materials? Or through a high degree of technology, visualized connectivity, visible autonomous control mechanisms, lightweight features, or virtuality? Is the renewed pleasure taken in the "good old (analog and mechanical) things" already a manifestation of a new sustainable/resilient attitude or a sentimental reminiscence of the past? Whatever the case may be, the symbolic character of products, just like their other two core attributes, must be created very carefully in order to produce durable products.

Holistic product design, i.e., the intelligent creation of these attributes and their intertwined border areas, not only results in holistically designed products, but also in consciously designed *product messages*.[7] Product messages lead to purchase decisions and influence user behavior. They are highly relevant for sustainable product design.

[7] The author analyzed these interdependencies in depth in her book titled "Product Messages: The Automobile as a Vehicle for Conveying Messages and as an Expression of One's Self." See Etezadzadeh (2008).

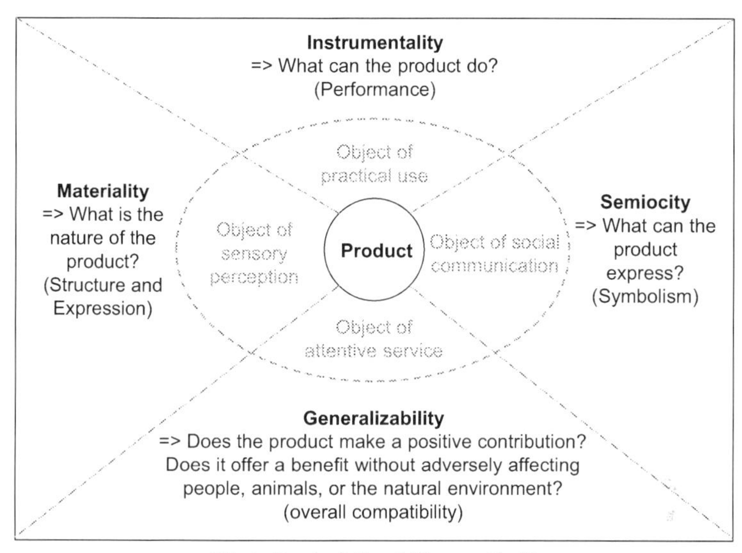

Fig. 4.3 Product attributes including the additional requirement to be generalizable. (Source: Based on Etezadzadeh (2008), p. 39)

In connection with product message research, the question arises as to what consumers do with the product? Because in accordance with transmitter and receiver logic, the product message is in fact shaped by the manufacturer but complemented by individuals. The urban design approach should go further and additionally analyze what the product does to people and our natural environment. For instance, to what extent does the product determine our behavior, cause people to be exploited, and consume resources? If these questions are taken seriously, product development becomes a responsible task. Sound product messages can definitely be useful and offer urban consumers desirable/sustainable messages for further use.

4.3 Generalizability

Just as we expanded the definition of the term sustainability/the municipal metagoal system in Sect. 2.3, we will likewise expand the definition of the attributes relevant for product development. In a globalized world with a steadily growing population but limited resources, the supplementary requirement for a product to be *generalizable* is a suitable approach for anchoring the sustainability obligation in the product development process (see Fig. 4.3).

Of course, this requirement poses an impossible task for numerous manufacturers of material goods. This obligation seems grotesque. However, it is necessary to realize that the living standard established in the industrial nations is not generalizable using currently available technologies, yet it is precisely what consumers in emerging countries strive to attain. It is doubtful that the future availability of abundant and practically free energy, sharing, and products stemming from a 3D printer can be considered a promise of salvation. Even in this case, we at least need to find solutions to current problems.

The first question that presents itself is what could constitute a generalizable product (possessing the objective character of an attentive servant) and whether generalizable products can even exist? These questions are difficult to answer. An organically farmed head of cabbage or a nearly maintenance-free, small-scale plant for producing renewable energy at least appear to harmonize with this ideal more than a technologically outdated SUV. Then we can go on to ask: If generalizable products existed, should consumers who can financially afford a lifestyle characterized by insouciant pleasures forego everything that is fun but not generalizable? And what about the quality of life? These questions are warranted, and no authority will be able to provide a universally valid answer. It is crucial, however, for consumers around the world to become aware of and have adequate information available to them so that they can become aware of what each of their actions as a consumer entails. They will then be able to decide for themselves whether they consider the consequences of their buying decision acceptable or not.

On the other hand, it does have advantages if product developers already address the issue of a product's generalizability during the design process. That would compel them to deal with questions regarding the product's purpose (do we need illuminated slippers?), design (are the required production processes ethical?), and materials used (will renewable resources be used?), etc. Putting this requirement on the agenda could be groundbreaking, especially in view of steadily growing cities and the requirements associated with their functionality.

Admittedly, these ideas may appear very unrealistic. So where do demand and reality overlap? With respect to material goods, we could perhaps agree that there simply would not be enough room in Tokyo for 38 million cars, for example. Yet mobility is a key element in our life and economic activities. Automobiles—in whatever form—will continue to exist, will help us overcome our physical limitations, and will continue to be like a partner and an extension of one's self for those who can afford it. Besides automobiles, we will therefore need additional, appealing mobility solutions for cities aimed at minimizing land use, air pollution, and noise emissions.

Having anticipated this trend, car manufacturers have expanded their core business. They produce cars with conventional and new engines, new types of construction and spatial concepts as well as cars with new functions and functionalities. At an early stage, they started to direct their attention to providing additional mobility options, which they are continuously improving and incorporating in new business models. Especially cities offer tremendous opportunities for enhancing old business models, but that would require a completely new way of thinking. Particularly in the premium segment, many approaches present themselves that would bolster the brand, promote brand loyalty, and attract new target groups as customers. This is how the automotive industry, which always had a role model character, is making advances towards fulfilling the requirement for its products to be generalizable. Other industries will have to follow the vanguard since their current value creation mechanisms will no longer work in the future as they have up to now.

This raises the question as to how virtual products meet the requirement to be generalizable. Is the partial relinquishment of materiality (e.g., through sharing) the way to achieve generalizable consumption? How do digital products and applications influence consumers? In the following chapters, we will address these questions and briefly discuss the issues associated with them.

Social Developments 5

In order to achieve successful business development[1] and design sustainable products, decision makers and corporate designers should understand what is going on in their environment, in their markets, and in the field of technological advances. In connection with her consulting activities, the author provides an easy-to-use tool that operationalizes this 360° view and makes it accessible to corporate industry and to medium-sized and small companies as well.

On the basis of this method, the author will in the following interlink some of the *mega trends*, trends, and developments pertaining to this study. In addition to many others, these topics are shaping urban society today and presumably will continue to do so in the years to come. Our analysis of the city as the subject of this study has therefore been expanded to include general social currents. We will outline the context of urban development and elaborate on urban product requirements, paying particular attention to the topics of sharing and trust, since sharing is currently being widely discussed and in part idealized, while the concept of trust appears to be undergoing considerable change. Without placing too much emphasis on developments occurring in industrial nations, the following discussion highlights trends that have emerged worldwide.

5.1 Ecology, the Economy, and Politics

Ecology

Society as a whole is showing an increased interest in preserving its natural environment. The public has come to understand the *finite nature of resources*. More and more people are becoming aware of the fact that the destruction of our natural

[1] cf. Etezadzadeh, C. (2009a), n.p.

C. Etezadzadeh, *Smart City – Future City?*, essentials,
DOI 10.1007/978-3-658-11017-8_5

resources is irreversible (e.g., with respect to fertile soil). A growing portion of the population thinks change is necessary, also and especially in view of *climate change*. Businesses are beginning to address this issue, not least because political requirements compel them to do so. However, the global community does not yet show any noticeable signs of implementing changes in order to stop these developments. Customers expect manufacturers to offer environmentally friendly products, but are not willing to accept significant price hikes.

The Economy

As regards the economy, it is becoming clear that global economic growth in its present form is based on the exploitation of our natural resources, at once jeopardizing many people's means of existence. This type of *destructive economic activity*, coupled with the exploitation of people as disenfranchised members of the labor force and the use of animals as objects serving production purposes, is being viewed ever more critically. This critical assessment of the current economic situation is underscored by the prevailing opinion that few people tend to benefit from the generated income.

The *polarization of wealth*[2], mentioned above has gained momentum through the decoupling of capital growth, the formidable concentration of power in the financial and economic sectors, structural changes, and unequal access to education, etc. This trend is increasingly manifesting itself in a critical analysis of the prevailing economic systems. *Globalization* is another contributing factor. It offers many advantages by providing opportunities for different cultures to share their knowledge, which is an aspect that all sides need to embrace and encourage. Yet it also forces businesses and their employees to compete internationally, poses a challenge to established structures, routines, and procedures, and calls into question the trove of knowledge gained through experience. Globalization—perceptibly—increases competitive pressure and gives rise to a demand for cost reductions and productivity gains through enhanced performance, efficiency measures, and cost savings, for which there is ostensibly "no alternative." Companies need to grow profitably in order to retain their competitive edge and protect themselves from takeovers.

The *financial and economic crisis* has helped to substantiate assertions made by critics of the system. Protagonists claim that self-absorbed mismanagement in the banking sector caused the crisis that for years has taken hold of the global eco-

[2] cf. OXFAM (2015), n.p. According to Oxfam, 1 % of the world's population will own more than half of global wealth in 2016. At the same time, 80 % of humanity will share 5.5 % of global wealth. One billion people will still live on less than $ 1.25 per day.

nomic system. It has destroyed numerous people's means of existence and, driven by monetary factors, the consequences of this crisis continue to result in the expropriation of the working population, while to outsiders nothing appears to have changed in the financial world. At the same time, *saturation effects* and the lack of any physical destruction of capital (e.g., through major acts of war in the industrial nations) have caused the global economic system to display a significantly slower growth rate than in the second half of the twentieth century. The business world is therefore pinning its hopes on future consumers in the emerging countries.

The various threats posed to one's very existence that are there for all to see give rise to growing *existential pressure*. It is necessary for people to partake in socio-economic development and to be spared from unemployment and a lack of prospects—factors that foment boredom, frustration, and hatred in so many places. Since the postwar period, children in industrial nations have for the most part been given the opportunity to develop the qualifications they need to pursue "successful gainful activities." In emerging countries a first or second generation is growing up that has similar opportunities in part. In both of these regions of the world, people are under pressure to succeed. In light of the overall situation outlined above, those opportunities alone are no longer sufficient to safeguard one's own existence with certainty. People around the world break under this pressure.

The global community appears to have moved closer together through the process of globalization, which has resulted in the *democratization of access* in many areas. More people have hopes of improving their standard of living and more people can afford things (thanks to minimum labor and logistics costs, among other aspects) that used to be considered out of reach. Nevertheless, it is obvious that this development, as described above, is not generalizable. It is interesting to note that these correlations have economic consequences, the ramifications of which we are perhaps not yet aware of. The availability of inexpensive "globalization products" makes things lose the value originally assigned to them (value deterioration due to lack of assignability). Since the negative effects of global production are not factored into the prices of products, today's consumers no longer know what the price of a product would have to be under healthy production conditions. This is the most direct kind of value deterioration.

Politics

On the whole, many people feel that their political interests are no longer being represented. Voters have the impression that politicians are unable to cope with the complex processes associated with the interconnected world and at the same time succumb to economic power structures. Necessary actions are not taken or do not have the anticipated effect. In keeping with the views propagated by the media, state apparatuses and supranational associations tend to work sluggishly, inefficiently, are incapable of reacting or reaching decisions, and occasionally focus more on their self-preservation and lobbyism than on serving the public interest. Citizens on every continent in the world condemn corruption and cronyism and criticize the way multinational corporations and high finance do business. Those attitudes do not represent a majority view, but they stimulate societal demand for more *transparency* and *participation.* Social unrest is brewing in countries that are strongly affected by such phenomena. This process regularly sets in as soon as people perceive an imbalance existing between the rights and obligations of the authorities and those of the public.

Especially citizens who are members of the *dwindling middle class* in industrial nations seem to be developing the feeling that they are no longer rewarded for "functioning" and for their willingness to function. They are furthermore afraid, and not without reason, that their social status is at risk. Even today, the readiness to conform and class-related virtues that characterized the twentieth century no longer have the effect they previously had: gaining recognition through good conduct. Faithfulness, loyalty, a sense of duty, commitment, and diligence are values that have been sacrificed for the sake of optimization. This leads to a *declining trust in rules*, a loss of orientation, disenchantment with politics, and dissatisfaction. These people long for tranquility, rest, and a certain amount of control.

Overall, modern man faces a great deal of change, insecurity, and a loss of control, which the Internet and other media additionally remind us of. *Fears* and *anxiety* about climate change, mounting injustice, fear of crime and violence, social unrest, terror and war, anxiety about the future in general, fear of uncontrollable technological advances, fear of epidemics and diseases, existential fear, fear of economic crises and unemployment as well as growing instability in personal affairs (partnerships and temporary families) increase people's subjective need for security to varying degrees depending on their environment.

The increasingly interconnected world allows differences between various lifeworlds to become visible, thus leading to migration. Driven by global imbalances and threatening situations existing in areas such as the environment, the economy, and the quality of life, *migratory movements* will continue to be on the rise (heterogenization of the population). Dealing with these trends by building higher fences will not be enough.[3] In cities in particular, many different people come into contact with each other in a crowded area. This togetherness needs to be shaped. Effective inclusion efforts based on education are therefore indispensable today and in the future.

[3] cf. ARD (2014). English video: cf. Euronews (2013).

5.2 Sociocultural Trends

One problem fueling all of the trends mentioned above is presumably the *economization of the lifeworld*. Spurred by digitalization, increasingly more things are being counted, measured, assessed, and compared in present-day life. Functionality, efficiency, effectiveness, usefulness, usability, and profitability are becoming more important. "Market imperatives have become the yardsticks of human thought and activity."[4] The saying that "there's no such thing as a free lunch" holds true more than ever nowadays, especially in the industrial nations.

Life that constantly focuses on opportunity costs is strenuous and results in a disintegration of the social fabric. Different contacts, different activities, and different value constructs determine a person's existence. People sleep less, eat faster, plan their life in order to make optimum use of the time available to them, and—either because they want to or are forced to—oust periods of carefree tranquility and idle leisure from their datebooks. They gradually become sicker, less creative, and above all hardly have any time for reflection.[5] That entails high costs and makes a society less resilient.

This situation gives rise to new needs: *simplification, ease, and comfort* are needed to cope with the harsh demands of everyday life. Technological and procedural innovations that serve to ease burdens are needed—but only innovations not requiring any further decisions to be made. Because key for those who can afford it is the *maximization of the quality of life*, an individually defined value that fundamentally shapes human endeavors but has never before been demanded to the extent that it is today.

The seeds of *individualization* originating in the industrial nations but sprouting around the globe form the basis of this trend. People have the prospect of acquiring the freedom to lead a self-reliant, self-fulfilled life. For those who have obtained it, that freedom means having to make decisions they did not have to make before. Many people furthermore perceive the freedom they have acquired as being a demand for continuous *self-optimization*. Both the decisions that have to be reached and the perceived need for self-optimization pose a challenge for people, and can become a burden.

Young adults are initially fully preoccupied with those tasks. Education and work become the focus of their attention, while plans to start a family have to be deferred in view of the aforementioned uncertainties, are discarded, or "revised" due to a separation. Today the iterative self-reorientation process continues well

[4] Heinzlmaier, B. (2012), p. 9.

[5] cf. Etezadzadeh, C. (2009b), n.p.

into old age (which evokes corresponding consumer desires, among other things). Altogether, these developments lead to a *pluralization of lifestyles* that exist side by side in cities.

Coupled with the changing *role of women* or because of financial dilemmas, *families* are emerging whose time management and daily routines can be termed demanding. Families need to coordinate the parents' jobs, the children's education which is considered important (insofar as it is affordable), and their upbringing, housekeeping chores, and the rest of family life (often with only one parent or in a patchwork format) with severe budgetary restrictions for the most part. Worldwide, families have needs, for instance in the areas of care, education, and relief from burdens, etc.

Demographic changes entail similar needs. Not all members of the aging population have families who can care for them. Therefore there is a need not only for affordable solutions for the care of the elderly and facilities that foster a sense of community, but also for decentralized senior-friendly solutions for health care services as well as accessible products and utility infrastructures.

A majority of *Generation Y* adolescents have grown up in family situations similar to those described above. Perhaps that experience is reflected in their insistence on a cooperative working environment based on partnership and the demand for work-life balance. Twentieth century middle-class role models, authoritarian hierarchies, and "analogous limitations" are things of the past for this generation. They are reputed to call things into question, to be well educated or to strive to receive a good education, and expect to have a fulfilled, meaningful personal and professional life. Perhaps Generation Y has adopted this attitude in response to the economization of life, or perhaps it is has already become the most natural thing in the world for them. Millennials are considered to be very self-confident, think of themselves as competent at an early age, and aspire to secure a job that entails responsibility. This goal becomes attainable thanks to digital entrepreneurship and can be regarded as a valuable addition to their realm of experience. It seems possible that information and communication technology and the millennials' affinity to those technologies will allow them to put their ideas into practice.

5.3 Sharing and Trust

Sharing

Sharing is an idea that could serve as a basis for digital entrepreneurship. It emerged as a result of digitalization (specifically Web 2.0, which we will discuss later on) and is regarded as an approach to finding a way to achieve more moderate consumption. What it means is that products are no longer purchased by everyone, but rather only by a few people who share them. That is efficient in many respects. Picking up on this idea, services are likewise subsumed under the term sharing. Thus, you can share an apartment or rent your whole apartment to someone, share a car ride with or lend your car to somebody, invite strangers to have dinner with you, or share your knowledge with them. Some of these services are offered free of charge, but usually they entail the payment of a fee. Sharing is like neighborly help on a global basis, although the sharing of consumer products is by nature predominantly a local phenomenon. However, the act of doing friends a favor can become a business venture. Sharing not only allows users to benefit by taking advantage of the solutions offered, which was initially the focal point of the idea, but also enables providers to generate profits. Sharing grants people market access with minimum barriers to entry. Everybody offers what they can, where and when they like. The basis for this is the Internet.

Overall, greater emphasis is again being placed on regionalism and individuality, and "smallness" is being created to counter globalization. Sharing evokes a sense of community, solidarity, and mutual help. It offers access and encourages urban production and the efficient use of resources. Depending on the business model, sharing thus generates resilience.

On this basis, the THINK and GROW incubator® was used to devise a model that can make cities all over the world more resilient. Mutual assistance, services, relief from burdens, and care will be the key to resilience. The author, as the innovator of the model, is promoting its propagation by inviting cities to exchange ideas with her, to address the topic, and roll out the model together with its residents in the interest of the city.

Of course, sharing also has potential disadvantages. The economization of sharing unfortunately entails the risk that aspiring small companies with a tendency to drift into a micro precariat situation, and to thus become less resilient, will inundate the market. Furthermore, companies using conventional business models rightly feel threatened by sharing offerings because the de-professionalization of competition undermines their efforts. Whereas conventional businesses must meet stringent requirements, sharing services are at present practically unregulated. This discrepancy needs to be resolved without depriving the sharing system of its great flexibility and inherent agility.

One question remaining to be answered is how sharing changes us. Do we become more sustainable by sharing? Will we be able to do without most of our possessions in the future? Does sharing mean that we will no longer feel attached to the things we own? Will we henceforth go through life with a backpack full of belongings and a device full of virtual possessions? Has the demonstration of status become obsolete? Lots of questions, and one answer: no. For one thing, specialists fear *rebound effects*. That means consumers could spend the money becoming available to them for other things or activities that consume resources. Or they could take a liking to the things they tested through sharing and subsequently want to own them themselves. Sharing is not suitable for esthetes, perfectionists, pedants, control freaks, or nostalgics who enjoy surrounding themselves with and caring for their cherished possessions, which lend them a sense of identity. And as to thinking in terms of status: Whether our products stem from a 3D printer or are of a purely virtual nature, whether we "share" part of our belongings or not, the joy of ownership, the love of an object and how it affects the formation of our sense of identity, and conspicuous consumption as described by Thorstein Veblen will serve to counter these trends.[6] Even in virtual worlds, people spend money so that their avatar is well dressed. The desire to display our status will persist and may therefore—just like the design requirements resulting from sharing—not be ignored in any urban design approach.

Trust

What enables us to suddenly share personal belongings with strangers who in the process may even become our "friends" in social networks or even good friends? Has our view of human beings or our behavior changed? At any rate, there is one element that helps make something like sharing possible: putative *transparency*.

Every Internet activity leaves behind a lasting trail. Our activities create an image of who we are. We have a *data twin* who not only fully documents what we do, but whose public persona is also constantly subject to scrutiny by others. Someone who borrows something and does not behave properly will get a bad rating; someone who makes stupid mistakes will reap a shitstorm. The Internet will not forget either of these things. That creates a highly visible *digital reputation*. On the one hand, this results in transparency; on the other hand, our data twin makes us vulnerable. Many different types of misuse ranging from mobbing, manipulation, and espionage to identity theft are conceivable. The perpetrators could be individuals, members of organized crime rings, institutions possessing data sovereignty, or even state authorities.

Whoever shies away from participating in online activities because of this threatening scenario will be left out. The ingenuous user thus generates data and

[6] cf. Etezadzadeh, C. (2008) and Fuhrer, U., Josephs, I. (1999).

trusts that nothing is likely to happen. Hence, the online community is based on a *culture of mandatory trust*. Each individual can merely decide which information he or she wishes to intentionally disclose. That decision requires knowledge that needs to be communicated, especially to the inexperienced.

Online participation is characterized by indiscretion, which therefore requires people to have a certain amount of blind confidence. This aspect is associated with technical advances in general and with digitalization in particular. Just think of the general terms and conditions that many people blindly accept again and again because it is practically impossible to actually read them and the numerous downloads and applications whose contents hardly anyone can precisely describe. We trust that the music we paid for today will still be on our virtual drive tomorrow. But what do we do if that is not the case?

In a way, digitalization thus also entails *disenfranchisement*. We have no idea how our devices work, often we cannot repair them, and some users are not even authorized to replace the rechargeable batteries in their smartphones by themselves. At the same time, our gadgets make us transparent for certain providers. This can be the telephone service provider, the smartphone manufacturer, the search engine we use, the app programmer, the anti-virus software provider, or our trusted online shop; they all know what you are doing—all the time. And usually they also know what you think, do not dare to ask, and are searching for.

Trust is the basis for partnerships, friendships, collaboration, and business deals. What exactly determines trustworthiness on the Internet? Photos, posts, clicks, likes, labels? Unfortunately, all of those things can be purchased. Some companies have specialized in building digital reputations professionally. Mechanisms need to be put in place to help us assess whom we are dealing with in virtual reality.

The topic of transparency remains unresolved for the time being. There is no sign of any substantial improvement in sight. The problem associated with this issue is that very few, mostly private sector businesses possess absolute data sovereignty. Glorified by the public, they eventually become hegemonic meta-nations that cannot be controlled, since they can do what only they are capable of doing. Users should be made aware of this situation including its inherent dangers. Politicians need to get involved in order to regulate these processes through international cooperation. Any change in this area affects us all. We are actually moving closer together. This aspect delivers an argument in favor of the requirement for products to be generalizable as presented above.

6 Digitalization

6.1 The Relevance of Digitalization

That brings us to a mega trend that has a key impact on our topics of interest: *digitalization*. A prerequisite for digitalization is *connectivity*, which requires the expansion of networks (cable-based and wireless broadband networks) and the widespread use of Web-enabled devices (PCs, portable computers, tablets, smartphones, etc.). Network rollout and the number of devices in use are rapidly increasing around the world. According to expert estimates, 40 % of the global population was already using the Internet at the end of 2014.[1] If growth rates remain unchanged, this will hold true for 50 % of the world's population in 2017.[2] This is made possible especially by cable-based broadband networks (711 million connections at the end of 2014) in conjunction with a progressive expansion of fiberglass networks and a global growth rate of 1.5 % per year (expansion is expensive) as well as mobile broadband networks.[3] While more than 6.9 billion mobile phone connections already exist today (2020: approx. 10.8 billion)[4], the number of smartphone users is projected to climb from 1.76 to 5.6 billion users by 2019.[5] Concomitantly, LTE market penetration is on a steep incline. In 2010, 612,000 people were using LTE, in 2011 that figure had already risen to 13.2 million, in 2012 to 100 million, and by 2016 the number of LTE users is expected to exceed 1 billion.[6]

[1] cf. ITU (2014), p. 5.

[2] cf. Broadband Commission (2014), p. 12.

[3] cf. Broadband Commission (2014), p. 18.

[4] cf. ITU (2014), p. 3. According to GSMA, this figure includes M2M communication. cf. GSMA (2014), p. 54.

[5] cf. Broadband Commission (2014), p. 20.

[6] cf. Lam, W. (2013), n.p.

C. Etezadzadeh, *Smart City – Future City?*, essentials,
DOI 10.1007/978-3-658-11017-8_6

Accordingly, more and more people have direct, high-speed Internet access both at home and when they are on the go. This serves as a basis for the digitalization of business and personal life and can lay the foundation for the breakthrough of "digital dominance."

For an increasing number of people, a growing part of their life is taking place on the Internet. Online shopping, social networks, entertainment (music, film and television, gaming, photo services, etc.), dating services and job markets, correspondence and telephony, banking and trading, professional content, access to information, and much more keeps users busy and devoted to the Internet. That is how the aforementioned data twin manifests itself, documenting the user's thoughts and actions and making the virtual community appear to resemble a threatening version of a village community, not only because it is larger, more heterogeneous, more widespread, more anonymous, and as a result latently more criminally inclined than its counterpart in reality, but also because it never forgets anything either. However, people can put themselves at the mercy of the public in the form of data disclosure to an even greater extent by using portable devices. Apart from smartphones and tablets that constantly register where the user is at any particular moment and are quite capable of underpinning that information visually and acoustically either intentionally or imperceptibly, other gadgets connected to the Internet such as sports activity trackers, smart watches, data eyeglasses, and other products gather additional personal data. This digital data collection is enhanced by the app-aided self-tracking trend,[7] which allows the user to become a quantified self. The objective of adherents of this trend is to acquire a variety of data (including number of steps, heart rate, training units, quality of sleep, moods, lifeworld effects, intestinal activity, etc.) usually for the purpose of self-optimization or self-awareness. That intimate information in the form of data is collected and interpreted via apps or shared with others to serve as an incentive. Advanced applications furthermore act as a digital coach who gives recommendations or issues warnings concerning the user's personal development. This trend is reinforced by a movement embraced by biohackers who for similar purposes use implants to let their bodies become one with technical devices. The "Tamagotchification" of the human condition may perhaps seem amusing. Yet in view of the data collection points and the potential misappropriation of the data collected there or with respect to the official use of such information and other performance-related data by (health) insurance companies, employers, or direct marketers, the collection of such a copious amount of data is unsettling. Optimization processes are essential for our self-preservation, but approaches such as the ones described above quell the

[7] cf. on the topic of self-tracking: Werle, K. (2014).

zest for life, job satisfaction, peace of mind, creativity, communication, the willingness to help, hinder thinking, and ultimately cost time. They cause stress and thus waste resources. Consequently, they are not resilient and hence not generalizable.

The always-on mentality is already so pronounced in some parts of society that a natural countertrend is currently emerging which pays homage to controlled abstention: Digital detox is (…) "a period of time during which a person refrains from using electronic devices such as smartphones or computers, regarded as an opportunity to reduce stress or focus on social interaction in the physical world."[8] … in order to be online again afterwards. Considering that many lifeworld topics have shifted to the Internet, the statement made by the Silicon Valley investor Marc Andreessen, which has become a well-known saying, is understandable: "Software is eating the world."[9] In his 2011 plea for *digitalization of business models*, he enumerates the many industries whose onetime leaders today have been replaced by software-based companies and he points out that this trend will continue relentlessly. The author is also of the opinion that companies all over the world have to react. Using the 360° analysis referred to above, it needs to be assessed to what extent a company's own business model can be digitalized or digitally upgraded and whether any digital threats will emerge in the environment of that particular industry. It must be analyzed which social developments affect a company's own business model and in which form customer needs should therefore be fulfilled in the future. As before, in the digital world it is necessary to tailor business models and products to fully meet customer needs in order to have a chance to succeed.

Companies that hitherto used a conventional business model will display a tendency to require fewer employees after switching to a digital business model and digitalized production processes. In any case, they will need employees that have different qualifications than the majority of the available labor force has today.

6.2 Development of the Internet from the Users' Perspective

The Internet is undergoing constant change too, of course. People most likely have different opinions regarding the classification of the development stages. To put it simply, from the users' perspective these stages can be described as follows:

[8] Oxford Dictionaries (n.d.) keyword: digital detox. http://www.oxforddictionaries.com/de/definition/englisch/digital-detox. Retrieved on 01/27/2015.

[9] Andreessen, M. (2011).

Web 1.0 users had PCs and inputted data into the Web, for instance, by creating files and static homepages containing data that rapidly became obsolete. The result was sort of like a photograph collection of the world—chronologically staggered, static images. People fed machines with data in order to find things faster, calculate faster, and to document and tidy up the real world. Typical applications included the use of reference books, websites, and search engines. Software programs such as those used to generate offers and calculations based on spreadsheets were used as part of a user's daily routine. Web 1.0 users were chiefly busy searching and reading, leaving an initial trail. This stage of the Internet was mainly about content. The Web enabled people to participate and, near the end of this stage, to access knowledge worldwide.

Nowadays, as we approach the end of the *Web 2.0* era, the use of PCs has become more seldom as portable devices become more popular and users increasingly access the Internet via mobile devices. Information can be shared by various, dynamic means (blogging, posting, chatting, etc.). People link static and dynamic information by entering data and acting on automatically generated suggestions. The result is a dynamicized partial image of the world and its processes corresponding to a loose collection of videos interspersed with real-time information. People create images, learn and teach the machines about existing correlations and interdependencies and which interdependencies are desirable under specific conditions and which ones are not. This is about coping with complexity and recognizing and understanding causal relationships and other forms of interplay. Typical applications include comparing offers, i.e., structuring massive amounts of data, editing wikis, blogging, and social networking. Scoring models are available to help people reach decisions, and data mining serves to improve hit rates in personalized advertising. Web 2.0 users find, read, and engage in equal measure. They realize that they are leaving a data shadow behind them that reveals quite a bit about them. This stage of the Internet is about linking information. Web 2.0 generates communication, lets the world appear smaller, and conveys a feeling of democratization of the ability to participate and of increasing transparency

In the coming years, Web 3.0 will be fully implemented. Users will have portable devices and high-speed mobile access to the Internet. In addition, systems embedded in devices, machines, objects, products, and living things, i.e., in basically everything, will generate data. Each thing and some living things will obtain personal access to the Internet including their own address. Furthermore, some things (e.g., production systems) will be able to self-activate and make decisions (e.g., take an order) depending on their status (e.g., their operational readiness). Machines will collect data and interview the things and living things addressed concerning their own condition and the condition of their environment. This will produce piles of data (big data). Anything and everything will be virtualized, giving rise to a "3D Internet" or an "Internet of everything." This will result in a parallelization of the real and virtual worlds, or a dynamic real-time image of the world. Self-optimizing and self-learning systems will integrate the data collected and start to identify patterns based on the large amount of data available, i.e., they will begin to understand their meaning. Humans as a source of error (e.g., in entering data) and analog process disruptions will progressively be eliminated. Systems will give recommendations for complex actions based on automatically optimized algorithms once developed by humans. They will process information in new forms, from new perspectives, and in new contexts, which means that future search engines will be able to answer highly complex questions autonomously, for example.[10] Big data will become smart data. Typical applications could include merging data to provide personalized information that can serve as a basis for research or for business models, for instance. Software programs will recommend actions based on complex analyses of globally available mass data comprising innumerable parameters. They will create personalized cancer therapies for patients, design databased urban planning projects, or establish business optimization measures for entire value creation chains. Web 3.0 users will be in one and the same context together with their data twins. Human decisions will therefore gradually become fathomable and predictable. It will be all about gaining insights. Mounds of raw data are meaningless and as incomprehensible to people as life itself. Machines will structure the available information, combine it in a variety of ways, and thus generate meaning and insights. People with an affinity to technology expect machines to help provide answers to unresolved questions, achieve optimization, and find solutions to make the world a better, more sustainable, and manageable place. The question is what will people do at the end of that phase?

Web 4.0 will not be long in coming either. The author is of the opinion that universal embedded access will be available by the (late) 2020s. Embedded systems in the form of miniature implants, and subsequently in the form of nanocomputers, could provide open-minded users with continuous system access, but that would make them more accessible to or even controllable by themselves and others. Experts assume that automated personal agents will communicate with users directly and in a humanized manner. The world and/or its processes could largely be controlled via the Internet or a few supercomputers, whether intentionally or not. Algorithms will be able to make decisions and implement them autonomously to a greater extent than today (comparable to automated high-frequency trading). The Internet would consequently have the potential to become a controlling parallel

[10] cf. Berners-Lee, T. (2009).

version of the world with a tendency to dominate. In this context, people would generate data, for the most part passively, that machines would collect, analyze, and interpret, and configure for solution processes that they would likewise implement. What scoring systems or automated securities trading systems were for Web 2.0, or what the recommended defense strategy or the industry 4.0 factory is to Web 3.0, could well be medical diagnosis completely made by autonomous machines that subsequently perform heart (tele)surgery for Web 4.0. Web 4.0 users could thus achieve maximum ease, but could potentially be influenced by their data twin or by technological implants, nanocomputers, etc. This stage of development would essentially revolve around the power issue.

All of this may sound utopian, yet these developments are already looming in current world affairs. Taking a look at developments in the area of artificial intelligence, the Law of Accelerating Returns, and the S-shaped curves of innovation cycles, as futurist and Google director of engineering Ray Kurzweil does,[11] and realizing what the power of exponential growth entails,—but that we still cling to our linear way of thinking,—we can assume that the next abrupt development stage is already on the horizon and that it will at least keep pace with the progress made in the past 20 years. One furthermore gets the impression that the moment at which it becomes necessary to shut down might come as a surprise.

Overall, society is becoming more vulnerable. Today, software manipulation and malware already cause economic and physical damage in the real world. Experts therefore demand digital resilience and the implementation of extensive security measures in this area. In order for society to understand the need for security and accept and respect the relevant measures that need to be taken, and to allow this issue to be incorporated in everyday politics, the public should be informed so that users become aware of the consequences of their actions, among other things.

6.3 The Significance of Digitalization for a Sustainable City

The many challenges facing cities have been described in the preceding chapters of this book (refer to Fig. 6.1). It will be impossible to accomplish the tasks required to tackle those challenges without the implementation of technological innovations, especially in big, fast-growing cities. Technological solutions, not as an end in and of themselves but rather as enablers, will allow cities to be functional. Mandatory requirements for convincing the city's residents of the need for the projected implementation of such innovations include digital resilience, data privacy,

[11] cf. Kurzweil, R. (2005).

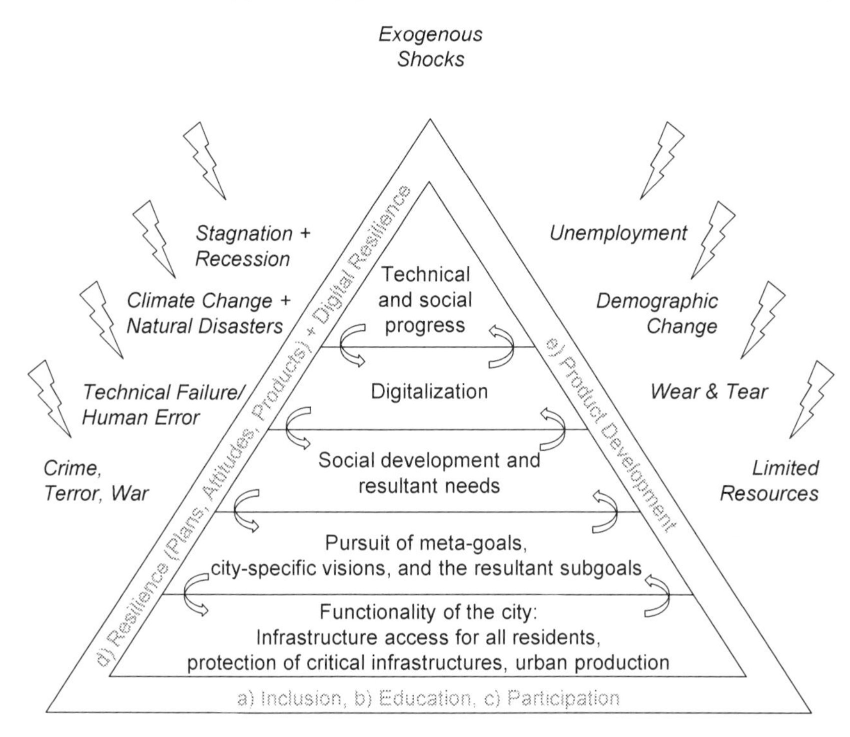

Fig. 6.1 Challenges facing cities. The various levels overlap, as indicated by arrows between the different stages. The security framework comprising **a** inclusion, **b** education, **c** participation, **d** resilience, and **e** product development protects the city from threats. (Source: author's own graphics)

nondiscriminatory data access, genuine functional value or relief from burdens through the innovations, and consideration of health aspects.

How does digitalization change a city? Cities need infrastructures that function effectively and efficiently. In particular, the energy, transportation, and security sectors, and afterwards the areas of health care and education, will probably undergo radical changes. Implementing those changes means that the city will have to be equipped with data collection points—as described above in connection with Web 3.0. Data collection will include measurements of the traffic volume on streets and capacity requirements at bus stops, streetlamps that collect environmental data, and waste containers that measure how full they are. The data gathered will be transmitted to the relevant authorities in real time, which should promptly trigger appropriate responses.

In addition, systems embedded in products, "smart homes" and smart phones, the stakeholders' activities, the use of online services, transactions, as well as movements and change will likewise produce data. The more extensively a city equips its infrastructures including its buildings, systems, machines, objects, and products with sensor networks and/or embedded systems and the more services that are available and used online, the more data will be continuously generated.

In order to plan the allocation of data collection points efficiently and to slightly curb the phenomenon of *ubiquitous computing*, the data collection points should be consolidated prior to implementation. That means a sensor would not measure merely one variable (e.g., brightness) but several values at the same time (e.g., also temperature, movement, etc.). A very inexpensive and very small system could thus at once provide not only one recipient but several interested parties with information about its condition and environment. Non-personalized data could be collected at a common data point such as an urban data platform to grant the various interested parties access to the data. All parties could avail themselves of the data they need for their own specific purposes and optimize their processes by using the available information. That would serve to ensure transparency and would not present any problems as long as the parties' objectives are socially compatible.

The vast amounts of data gathered this way are initially not very useful given their massive volume. This is referred to as big data. Algorithms are needed to transform mass data into informative data. Algorithms process piles of data and convert them into information. Based on the data that are continuously collected, past or present processes can be depicted, patterns identified, and future processes predicted. Patterns provide informative data, or smart data.

The availability of the quantity of data described above could be utilized to optimize water pipe pressure control, bus fleet deployment, or the allocation of police patrol cars before, after, and during soccer finals, for example. Data can serve to enhance business models, optimize advertisements, and promote or enable entrepreneurship. Prerequisites for this include the availability of reliable broadband networks that are capable of securely transmitting large amounts of data in real time at high transmission rates and the use of high-performance data processing centers or cloud computing. That way very small embedded systems consisting of microcontrollers, sensors, actors, identifiers, and communication systems, or some of those elements can change the options available for managing a city.[12]

[12] cf. BITKOM, IAO (2014).

In centers for security and safety, a real-time image of the city in conjunction with video surveillance of hot spots can serve to manage effective and coordinated actions. Whatever works in the city will work just as well at the city's manufacturing sites. Production at these sites is based on the "Industry 4.0" principle.[13] Self-learning, communicating, and self-regulating machines will optimize production processes across entire value creation chains as they fulfill orders in a coordinated and autonomous manner. Intelligent materials will make their way into intelligent products that are capable of buttressing the urban design approach during all product stages until they are recycled. These processes will greatly ease the burden on humans, but also replace them. Such *cyber-physical* systems prevent human errors from occurring but lack human rationality and common sense.

Digitalization can make cities more transparent, safer, and more functional, yet more vulnerable to sabotage as well. Digital resilience, data privacy, and information security will therefore become more important. People will have to use their judgement to assess the extent to which we need technologization, automation, digitalization, interconnections, and decentralization in our cities. Social developments and the ensuing societal needs, the desire to keep up with technological advances, and the digital revolution will most likely push these boundaries.

[13] cf. ibid.

The City of the Future

7

7.1 The Energy Sector (Electricity)

Considering that technological advances will play a major role in modern-day life and in a sustainable, digitalized city, the question arises as to how to meet the accompanying demand for electricity. Overall improvements in energy efficiency and a systematic reduction or even avoidance of power consumption as the only available options will not suffice. Renewable and abundantly available energy will form the basis for the digital revolution, or the third industrial revolution. It will not be possible to implement a sustainable, digital city—let us call it a smart city—without a revolutionized energy sector. This will require an urban energy revolution based on the implementation of a *smart grid*. Figure 7.1 shows a schematic diagram of the elements of an urban, almost decarbonized energy sector.

After the energy revolution, power generation will be decentralized. Consumers (e.g., households, public facilities, businesses) will largely produce energy themselves. Decentralized power plants operated by the municipal utility company will supplement those energy market structures to supply power to smaller consumer units that do not produce their own energy (e.g., old apartment buildings, streets, urban districts). All power supply units will operate completely on the basis of renewable energy.

The municipal utility company will keep flexible standby plants in reserve that can supply electricity at short notice in a resource-efficient manner in the event of a special or emergency situation or if a sufficient supply of renewable energy is not available. The reserve capacity serves to ensure a reliable supply of power, on the one hand, and helps utility operators respond to demand (load management) and manage the electric power grid (network control), on the other.

Households (smart homes) will thus generally be prosumers and will be able to produce, store, consume, and not only additionally purchase, but also feed

C. Etezadzadeh, *Smart City – Future City?*, essentials,
DOI 10.1007/978-3-658-11017-8_7

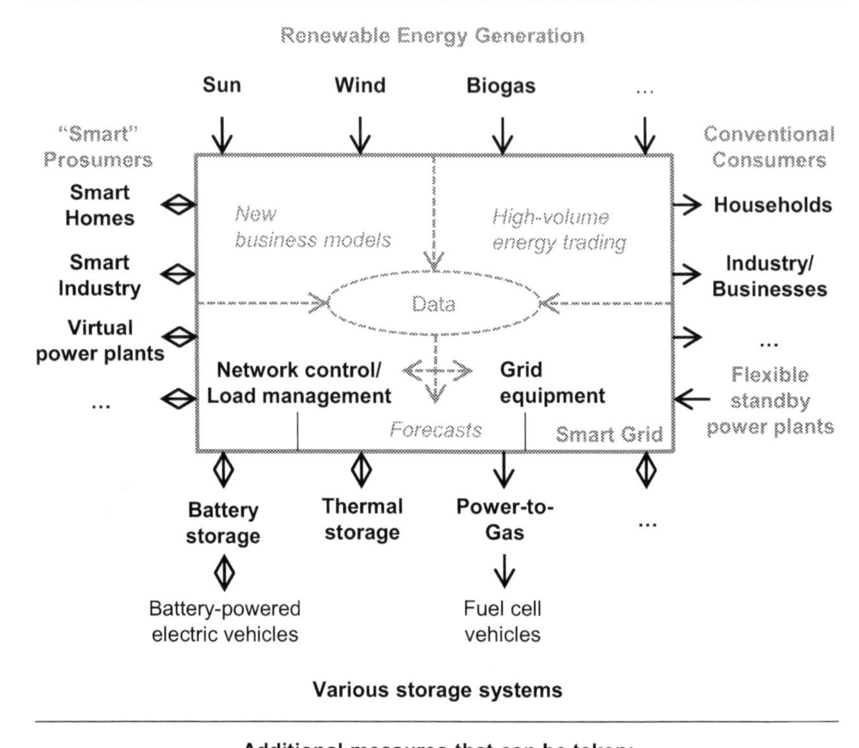

Additional measures that can be taken:

Reduce/Avoid power consumption

Smart Meter Roll out

Improve energy efficiency incl. retrofitting

Fig. 7.1 Power supply (electricity) in a sustainable, digitalized city. (Source: author's own graphics)

electricity into the grid by means of bidirectional power networks. The same applies to the other consumers that—just like households—will adapt their power consumption and demand according to availability and relevant price levels without having to make any major sacrifices. The production or storage capacities of several small and/or local power supply units will be pooled to create virtual power plants. That way they can be integrated in the power supply system in a manner that is appropriate to the grid and flexibly provide significant amounts of electricity (balancing energy) depending on demand.

To help ensure consistent availability and reliability of electricity in spite of the volatile generation of energy from renewable resources, a variety of options are available to create a reserve storage capacity. This includes battery storage systems (which are gradually becoming commercially viable), thermal storage systems, power-to-gas solutions, and other storage technologies that will become available in the future. Fuel cell vehicles will consume the hydrogen generated by P2G systems and larger fleets of battery-powered electric vehicles will serve to store electricity temporarily and to offset peak loads. Electric vehicles owned by households equipped with integrated energy management systems will initially augment the home's own optimized power supply, and later on, when pooled, that of the urban districts, and so on.

The various electricity producers, storage systems, consumers, and grid equipment will be interconnected and managed by linking them via communication networks and power grids to create a smart grid. The power grids will be adapted to meet new daunting challenges related to load, volatility, and bidirectionality. Supplementary communication networks will serve to coordinate erratic power production and unpredictable consumption to help improve grid stability.

Universally installed smart meters and the smart grid will produce enormous amounts of data. Analysis of the metadata can provide useful demand forecasts, which together with other forecast data (e.g., weather, capacities, plant availability, etc.) will allow the grid operator and thus the urban consumption and production community to operate the power grid efficiently, while helping to ensure that it has a long service life. Some of these data could be incorporated in the municipal data pool. However, the urban community should definitely be consulted as to whether and to what extent data may be disclosed.

In this future scenario, completely new value creation chains present themselves for energy suppliers. Customized solutions for their customers who are meanwhile self-producers as well as new services will expand or replace their range of activities (in part in the form of collaboration across industries). New market roles such as that of a service provider in the field of power trading will emerge, requiring the relevant products to be made available at an early stage.

Several of the author's current studies focus on the reorientation of energy suppliers.[1] During the past years, one of her chief recommendations was to "bring the customers to the business" in order be able to better understand their needs and serve them better. Since dedicated customer orientation has been incorporated in the CIP (continuous improvement process), it is now imperative to "bring the

[1] Public utilities in particular play a key role due to their interface function in an urban context.

business to the customer" in order to maintain lasting customer relationships and defend them against new stakeholders despite the forthcoming decentralization.

In this context, as electric vehicles begin to play a new role, new opportunities will present themselves for car manufacturers as well. In conjunction with energy management systems and equipped with bidirectional charging systems, vehicles will acquire a new social relevance. They will evolve from cost-incurring self-extensions into versatile enablers that will in part pay off. This will offer energy suppliers, automobile manufacturers, and other industries new opportunities to co-operate, which should be assessed ad hoc to be considered strategically.

7.2 Structure and Features of Smart City 2.0

We have now become aware of some of the key challenges facing cities, gained insights into their society, discussed their digitalization, and have come to realize what drives them. In the following, we will incorporate those insights in the specification of requirements for a smart city in terms of a sustainable, digitalized city. Numerous approaches for accomplishing this have been published in the past. Taking that preliminary research into account and addressing the expanded requirements defined in this study, we will describe the features of *Smart City 2.0* below.

Figure 7.2 illustrates our results. It shows a schematic diagram of the structure of a Smart City 2.0, which we will use as a basis for defining a list of features. The city's structure can be described by seven levels of *enablers*:

1. Natural basis
2. Urban stakeholders and their contributions
3. Integrated municipal government and urban governance
4. Goals and visions
5. Infrastructures
6. Level of information and communication technology
7. Resilience

Re 1. The *climate, the natural environment, and the limited resources of a city* are defined as natural enablers in this context. They form the necessary basis for any type of urban life and must therefore be protected and preserved accordingly. Smart City 2.0 ultimately aims to achieve controlled extraction and recycling of material to allow natural regeneration as well as the complete elimination of harmful emissions.

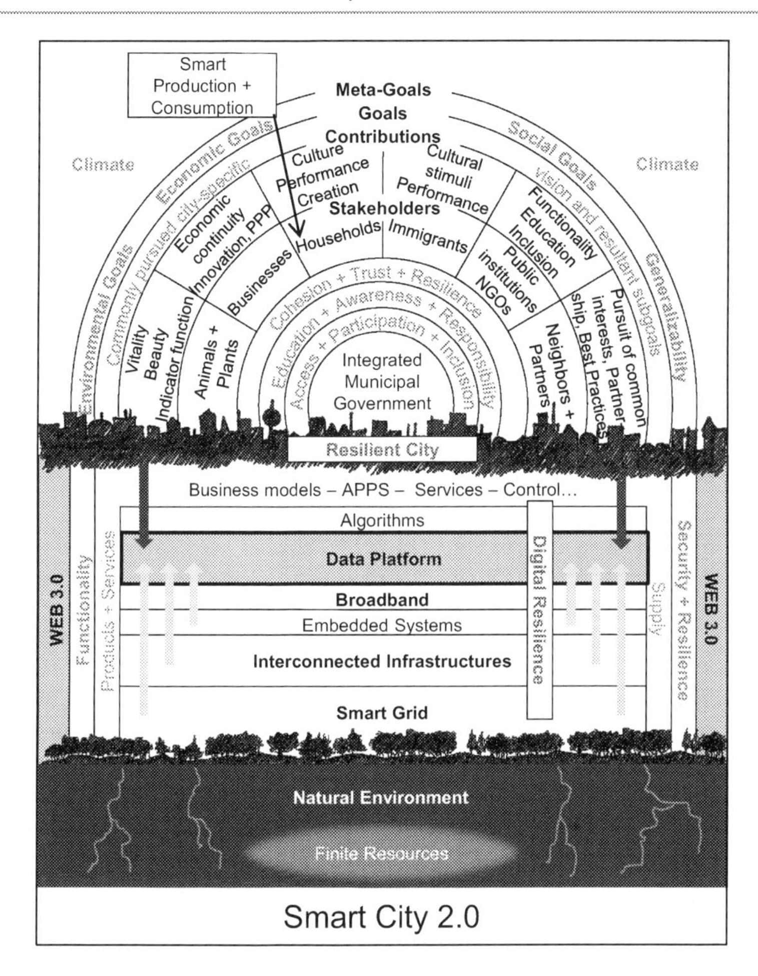

Fig. 7.2 Structure of Smart City 2.0. (Source: author's own graphics)

Re 2. The second level describes the *human and active stakeholders*. Urbanites and their interests are the source of urban life. They make a city come alive. They create the city and represent the city. Apart from maintaining the natural enablers, it is therefore just as important for the municipal government to act in the interest of the city's residents. All governmental activities should center on the residents' potentially conflicting needs and encourage the urbanites' manifold contributions as well as their access, inclusion, and participation (including the promotion of economic development). Educational measures should be implemented for the stakeholders to generate awareness and a sense of responsibility. This forms the basis for social cohesion that can bring forth a trusting and resilient community ("smart citizens") committed to attaining common goals (e.g., environmental protection, moderate use of technology, data privacy, etc.). Smart citizens are aware of their dependencies and multifaceted responsibility. They therefore produce and consume primarily generalizable products.

Re 3. The third level has not yet been addressed. It stands for *integrated municipal government* whose activities are guided by the concept of *urban governance*. Urban governance denotes a continuous process. It serves the purpose of integrating the various plans and activities of the formal and informal structures of the public, private, and civilian sectors to manage urban affairs and to ultimately achieve a communal ability to act.[2] The municipal government represents the interests of the city on all levels of the national government and establishes its political and regulatory framework. It is responsible for managing the municipality and reaching investment decisions. In doing so, it actively participates in shaping all five levels of enablers in a coordinated manner as well as in the interest of the city's goals and an integrated urban development plan.

An integrated municipal government has adapted to meet structural, organizational, and procedural demands. It strives to achieve efficiency, transparency, and sustainability and advocates the principle of subsidiarity. Transparency regarding its performance is a given as is cross-departmental, intersectoral, and interdisciplinary cooperation. It has the ability to reach decisions and act. It continuously seeks to cut red tape and further the digitalization of administrative actions, participation processes, incentives for good citizenship, economic development, and public-private partnerships.[3]

Re 4. Together with the city's residents, the municipal government develops a city-specific, sustainable vision based on the city's four meta-goals. Specific goals for action result from this vision, which the stakeholders feel committed to since it was jointly developed. This gives rise to the urban goal system referred to here as the level of *goals and visions*.

Re 5. The fifth level concerns the *infrastructures*. It consists of appropriately dimensioned, flexible, interconnected, integrated, resource-efficient infrastructures that operate efficiently (in this case meaning all urban infrastructures even if they overlap with other levels of enablers). As part of an integrated urban development plan, the intersectoral interconnection of infrastructures enables synergy

[2] cf. UN-HABITAT (2002), n.p.

[3] cf. UN-HABITAT (2002), n.p. and GDRC (n.d.), n.p.

achievement, maximum availability of information, comprehensive process control, and the realization of economies of scale, scope, and density. Thanks to the urban energy revolution, a sufficient supply of energy from renewable sources is available to the city.

Re 6. *Information and communication technology* is a separate level spanning all municipal levels of enablers. It enables the integration of infrastructures and numerous urban processes. It benefits all levels of enablers except the natural ones in making their contributions and achieving their goals. An urban platform could potentially provide all city residents with access to the data collected in many places, while ensuring data privacy and maximum data security. However, the use of software in Smart City 2.0 is subject to constraints. Technology will be prevented from dominating human life. Decisions will continue to be based on human judgement and rationality.

Re 7. Due to the necessary digitalization as well as natural and anthropogenic threats, the city becomes vulnerable. It strives to maintain its functionality and ensure the security of its citizens through digital resilience and *resilience* in the area of its infrastructures as well as through diversification of urban production (partial urban autonomy, urban farming, etc.), through resilience in process and product design, and through a far-reaching cultural change in the city.

7.3 What is Smart City 2.0?

Smart City 2.0 can therefore be described as follows:

It is a community aimed at individual and urban (self-) preservation comprising all groups of human urban stakeholders. Their behavior (including production and consumption) is completely geared to the urban goal system jointly developed by all of them on the basis of the city's meta-goals (sustainability and generalizability). They are committed to their diverse community goals, champion their sovereignty as consumers, residents, and humans, as well as the protection of their city's natural environment and wildlife. To achieve this, they employ technical facilities to a great extent, but do not allow technology to expand uncontrollably, dominate urban life, or acquire decision-making authority.

The human stakeholders have mutually agreed to maintain their city's functionality and the safety of its residents at all times. Interconnected and integrated infrastructures are understood to form the basis for that functionality, are treated accordingly, and effectively satisfy the urbanites' consumption needs. A comprehensive culture of resilience has been achieved for the benefit of the city's functionality.

Through education, the stakeholders have developed an awareness of the overall urban situation and a sense of responsibility for dealing with urban threats. Access, inclusion, and participation mobilize the stakeholders. This results in social cohesion, trust, and greater security. The integrated municipal government encourages these developments and the contributions made by the stakeholders (while promoting the economy as well) and acts according to the principle of urban governance. Through coordinated actions and informed investment decisions, the municipal government oversees urban development in the interest of the urban goal system and an integrated urban development plan.

7.4 Conclusions

What does all of this tell us? Let us first take a look at the economy. Cities need businesses that fully understand the overall situation of the city and the challenges it faces and develop solutions for urban needs that satisfy the product requirements described above to the maximum extent possible. On the one hand, the product offerings range from robust basic infrastructures for cities struggling with the most fundamental, yet dramatic supply and disposal problems up to integrated high-technology solutions for megacities with mega processes. On the other hand, they range from micro business models that generate smallness and resilience up to killer applications. Those two approaches can serve to counteract persistent stagnation and rising unemployment. This type of economic activity necessitates a global energy revolution as set out above as well as the creation of the conditions for implementing such a transition.

New business models and new value creation chains will emerge on the basis of intersectoral cooperation. Due to urban economic restrictions those business models will have to include a new financing approach. That may require new profitability criteria, long-term investments, and in part a new understanding of growth. At the same time, new stakeholders and digitalization will undermine existing structures. The complexity and interdisciplinary nature of the projects will pose new challenges for all those involved. That will require systemic methodological competence, the willingness and ability to think synthetically, and multilateral interface management. Cities will need to prepare for these requirements as regards their structure, organization, regulations, and procedures. Despite all of these obstacles, cities must be viewed as key markets of the future that possess great value creation potential and need to be developed.

In the sociocultural area, the role of the urbanites has turned out to be quite demanding: they need to be educated, assume responsibility, and consume more prudently. Yet the majority of urbanites are already completely busy with their daily struggle to survive. They are far from being able to assume additional responsibility and make choices as consumers. The more well-to-do, however, might not be interested in that kind of sovereignty. After all, individual responsibility that is made possible by the Internet costs time and requires effort.

The only feasible way to overcome those obstacles is through education. It offers an opportunity for self-preservation and is thus a basic prerequisite for empathy. Education furthers people's cognitive interest and their ability to process knowledge gained through experience, thus enabling them to develop their power of judgement. Education can help people learn what is good or harmful for them and their environment. Differentiated perception is perhaps a general twenty-first century requirement. Because today we need to decide how we wish to live in the immediate future.

Cities as smaller and larger units have great potential for changing from within. They are in a position to launch reversal processes—in spite of challenges that appear to be insurmountable in part (such as the irreversible destruction of our natural environment or the continuous proliferation of slums) and in spite of the given economic restrictions. In this context, the laws of exponential growth also apply. Intelligent people who think and (loosely based on Kant) have the courage to use their own reason are a city's most valuable tool. They serve as a basis for the sensible design of livable cities and the preservation of our natural environment.

Please allow me to close this analysis with a personal statement:

Rationally, we realize that unlimited economic growth and the exploitation of nature as practiced today is not possible in a world with limited resources.[4] In order for us to not only rationally understand, but also to become emotionally aware of why we should protect natural diversity and all that is vulnerable, we would need to once again bond to all things living[5] and develop a cognitive interest in them.

We cannot do justice to the other being without exploring its essential nature. That kind of attitude can result in reasonable behavior guided by empathy. Empathy with oppressed human beings, with animals that are abused and used for mass production purposes, and with our ailing environment. That would be desirable, since empathy and the right balance (also with respect to ourselves) are perhaps the prerequisites for Smartness 2.0.

Let all of us together help to ensure that good times await us.

---

[4] cf. Capra, F. (2010).

[5] cf. Capra, F. (2010).

What are the Take-Home Messages of This *Essential*?

- Cities are above all people and need the natural environment as a basis of existence
- Functionality and resilience are the top urban objectives
- Technical advances and an urban energy transition are indispensable enablers for reaching those goals
- Education is an essential requirement for viable cities of the future
- Smart City 2.0 can be livable and is a future market posing new challenges

C. Etezadzadeh, *Smart City – Future City?*, essentials,
DOI 10.1007/978-3-658-11017-8

References

Andreessen M (2011, August 20) Why software is eating the world. Wall St J. http://www.wsj.com/articles/SB10001424053111903480904576512250915629460. Accessed 27 Jan 2015

ARD (2014, March 23) Spanien/Marokko: Der tödliche Zaun von Melilla [Video]. http://www.daserste.de/information/politik-weltgeschehen/weltspiegel/videos/spanien-marokko-der-toedliche-zaun-von-melilla-100.html. Accessed 16 Jan 2015

Bähr J (2011a) Einführung in die Urbanisierung, in: Online-Handbuch Demografie. Berlin-Institut für Bevölkerung und Entwicklung. http://www.berlin-institut.org/online-handbuchdemografie/bevoelkerungsdynamik/auswirkungen/urbanisierung.html. Accessed 06 Jan 2015

Bähr J (2011b) Ursachen für Urbanisierung, in: Online-Handbuch Demografie. Berlin-Institut für Bevölkerung und Entwicklung. http://www.berlin-institut.org/online-handbuch-demografie/bevoelkerungsdynamik/auswirkungen/urbanisierung.html. Accessed 06 Jan 2015

Berners-Lee T (2009, February) The next web—TED2009. http://www.ted.com/talks/tim_berners_lee_on_the_next_web. Accessed 30 Jan 2015

Broadband Commission for Digital Development (2014) The state of broadband 2014: broadband for all. http://www.broadbandcommission.org/Documents/reports/bb-annual-report2014.pdf. Accessed 27 Jan 2015

Bundesministerium des Innern—BMI (2009) Nationale Strategie zum Schutz kritischer Infrastrukturen. http://www.bmi.bund.de/cae/servlet/contentblob/544770/publicationFile/27031/kritis.pdf. Accessed 09 Jan 2015

Bundesverband Informationswirtschaft, Telekommunikation und neue Medien e. V.—BITKOM, Fraunhofer-Institut für Arbeitswirtschaft und Organisation IAO (ed) (2014) Industrie 4.0—Volkswirtschaftliches Potenzial für Deutschland. http://www.bitkom.org/files/documents/Studie_Industrie_4.0.pdf. Accessed 03 Feb 2015

Capra F (2010) Interview im Rahmen der 8. Schweizer Biennale zu Wissenschaft, Technik und Ästhetik. http://www.art-tv.ch/5213-0-Biennale-Luzern-Fritjof-Capra.html. Accessed 15 Feb 2015

Dateline SBS (2011, September 25) E-Waste Hell. https://www.youtube.com/watch?v=dd_ZttK3PuM. Accessed 14 Jan 2015

Deutsche Akademie der Technikwissenschaften—Acatech (n.d.) Dossier Sicherheit. http://www.acatech.de/sicherheit. Accessed 10 Jan 2015

C. Etezadzadeh, *Smart City – Future City?*, essentials,
DOI 10.1007/978-3-658-11017-8

Etezadzadeh C (2008) Produktbotschaften (Product messages)—Das Auto als Botschaftsvehikel und Ausdruck meines Selbst. Dissertation, Witten-Herdecke

Etezadzadeh C (2009a) Utilize chance!—What is business development? http://www.thinkandgrow.de/tag/index.php/en/what-is-business-development. Accessed 19 Jan 2015

Etezadzadeh C (2009b) Holistic thinking for sustainable growth. http://www.thinkandgrow.de/tag/index.php/en/why-brtag-tag-consult. Accessed 16 Jan 2015

Euronews (2013, December 9) Melilla: the Spanish enclave that has become the back-door to Europe. https://www.youtube.com/watch?v=ZHj58hYdhMg. Accessed 16 Jan 2015

Europäische Kommission, Generaldirektion Regionalpolitik—COM GD REGIO (2011) Städte von morgen—Herausforderungen, Visionen, Wege nach vorn. Amt für Veröffentlichungen der Europäischen Union, Luxemburg

Fuhrer U, Josephs I (1999) Persönliche Objekte, Identität und Entwicklung. Vandenhoeck und Ruprecht, Göttingen

Gabler Wirtschaftslexikon (n.d.) Stichwort: Infrastruktur. http://wirtschaftslexikon.gabler.de/Archiv/54903/infrastruktur-v9.html. Accessed 08 Jan 2015

GDRC Programme on Urban Governance (n.d.) Some attributes on Urban governance and cities. http://www.gdrc.org/u-gov/good-governance.html. Accessed 10 Feb 2015

GSMA (2014) The mobile economy 2014. http://www.gsmamobileeconomy.com/GSMA_ME_Report_2014_R2_WEB.pdf. Accessed 27 Jan 2015

Heinzlmaier B (2012) Jugendkulturen in Zeiten von Ökonomisierung und Moralverlust. http://www.fgoe.org/veranstaltungen/fgoe-konferenzen-und-tagungen/archiv/was-kann-gesundheitsfordernde-schule-verandern/Prasentation%20Jugend%20und%20Zeitgeist_14062012.pdf. Accessed 16 Jan 2014

International Telecommunication Union—ITU (2014) ICT fact and figs. 2014. http://www.itu.int/en/ITU-D/Statistics/Documents/facts/ICTFactsFigures2014-e.pdf. Accessed 27 Jan 2015

Kurzweil R (2005, February) The accelerating power of technology—TED2005. http://www.ted.com/talks/ray_kurzweil_on_how_technology_will_transform_us. Accessed 30 Jan 2015

Lam W (2013, January 22) Global LTE subscribers set to more than double in 2013 and exceed 100 Million. https://technology.ihs.com/419630/. Accessed 27 Jan 2015

Lexikon der Nachhaltigkeit (n.d.) Brundtland Bericht, 1987. http://www.nachhaltigkeit.info/artikel/brundtland_report_563.htm. Accessed 06 Jan 2015

OXFAM international (2015, January 19) Richest 1% will own more than all the rest by 2016. http://www.oxfam.org/en/pressroom/pressreleases/2015-01-19/richest-1-will-own-more-all-rest-2016. Accessed 19 Jan 2015

Oxford Dictionaries (n.d.) Keyword: digital detox. http://www.oxforddictionaries.com/de/definition/englisch/digital-detox. Accessed 27 Jan 2015

Pederson P, Dudenhoeffer D, Hartley S, Permann M (2006) Critical infrastructure interdependency modeling—A survey of U.S. and International Research. http://www.inl.gov/technicalpublications/Documents/3489532.pdf. Accessed 08 Jan 2015

Rat für Nachhaltige Entwicklung (n.d.) Was ist Nachhaltigkeit? http://www.nachhaltigkeitsrat.de/nachhaltigkeit. Accessed 06 Jan 2015

Revi A, Satterthwaite DE (2014) Urban areas, in: Climate Change 2014: Impacts, Adaptation, and Vulnerability. Part A: Global and Sectoral Aspects. Contribution of Working Group II to the Fifth Assessment Report of the Intergovernmental Panel on Climate Change. Cambridge University Press, Cambridge (UK), New York, 535–612

Schott D (2006) Wege zur vernetzten Stadt—technische Infrastruktur in der Stadt aus historischer Perspektive. Inf Raumentwickl 5(2006):249–257

Siemens AG (2006) Megacities und ihre Herausforderungen. http://www.siemens.com/entry/cc/features/urbanization_development/de/de/pdf/study_megacities_de.pdf. Accessed 08 Jan 2015

UN-HABITAT (2002) Principles of Urban governance. http://ww2.unhabitat.org/campaigns/governance/Principles.asp. Accessed 10 Feb 2015

United Nations, Department of Economic and Social Affairs UN/DESA, Population Division (2002) World urbanization prospects: The 2001 revision. United Nations, New York

United Nations, Department of Economic and Social Affairs UN/DESA, Population Division (2012) World urbanization prospects: The 2011 revision. United Nations, New York

United Nations, Department of Economic and Social Affairs—UN/DESA (2013a) World economic and social survey 2013. Sustainable development challenges. Chapter III, towards sustainable cities. United Nations, New York

United Nations, Department of Economic and Social Affairs—UN/DESA, Population Division (2013b) World population prospects: The 2012 revision, vol 1. United Nations, New York

United Nations, Department of Economic and Social Affairs UN/DESA, Population Division (2014) World urbanization prospects: The 2014 revision. United Nations, New York

WDR (2012, August 24) Agbogbloshie—Elektroschrott in Ghana bei WDR Planet Wissen. http://www.youtube.com/watch?v=qqYDWbVg2yw. Accessed 14 Jan 2015

Werle K (2014, April 17) Self-Tracking für Manager—Blöd, dass der Körper keinen USB-Anschluss hat. http://www.spiegel.de/karriere/berufsleben/self-tracking-im-job-die-besten-self-tracking-apps-fuer-manager-a-964940.html. Accessed 27 Jan 2015

ZVEI—Zentralverband Elektrotechnik- und Elektronikindustrie e. V. (ed) (2010) Integrated technology roadmap automation 2020 + Megacities. Frankfurt

Printed in the United States
By Bookmasters